2nd International Conference on Computer, Remote Sensing and Aerospace (CRSA 2022)

Journal of Physics: Conference Series Volume 2410

Nagoya, Japan
26-28 August 2022

ISBN: 978-1-7138-6549-0
ISSN: 1742-6588

Printed from e-media with permission by:

Curran Associates, Inc.
57 Morehouse Lane
Red Hook, NY 12571

Some format issues inherent in the e-media version may also appear in this print version.

Institute of Physics
Dirac House, Temple Back
Bristol BS1 6BE UK

Phone: 44 1 17 929 7481
Fax: 44 1 17 920 0979

techtracking@iop.org

Additional copies of this publication are available from:

Curran Associates, Inc.
57 Morehouse Lane
Red Hook, NY 12571 USA
Phone: 845-758-0400
Fax: 845-758-2633
Email: curran@proceedings.com
Web: www.proceedings.com

TABLE OF CONTENTS

Preface

The 2nd International Conference on Computers, Remote Sensing and Aerospace (CRSA 2022) was successfully held in Nagoya, Japan, on August 26-28, 2022. Due to the impact of COVID-19, it was changed to be held online. The conference mainly focused on the latest researches on Computer, Remote Sensing and Aerospace, gathering experts, scholars, and researchers in these fields from all over the world to share research achievements, explore hot issues, and exchange experience and new ideas.

In the conference, we were greatly honored to have both Prof. Yanguo Jing from Coventry University, UK and Prof. Jiang Zhu from Tokyo Institute of Technology, Japan to serve as our Conference Chairs. 100 scholars around the world took part in the conference. Divided into three parts, the conference agenda included keynote speeches, oral presentations and online Q&A discussion. Primarily, the keynote speakers were each allocated 30-45 minutes to address their speeches. Then in the next part, oral presentations, the excellent papers we had selected were presented by their authors one by one.

Three distinguished professors were invited to hold keynote speeches during the conference. Among them, Prof. Yanguo Jing from Coventry University, UK performed a thought-provoking speech on Artificial Intelligence Applications in Sports Performance. Although Artificial Intelligence (AI) are being used to create new values and opportunities in business and society, there are many challenges in developing these AI applications. In the speech, Yanguo shared his own journey of use artificial intelligence and machine learning to monitor sports performance, and evaluated their effectiveness in assessing energy expenditure during sport exercise. What's more, Assoc. Prof. Thippa Reddy Gadekallu from Vellore Institute of Technology, India addressed a speech: Federated Learning for Healthcare. He introduced the applications of FL for healthcare informatics, the need for FL in healthcare domain, the fundamentals of FL, and the major motivations behind FL for healthcare applications, etc. And some lessons learnt, challenges, as well as future directions were highlighted. Their insightful speeches had triggered heated discussion. And every participant praised this conference for disseminating useful and insightful knowledge.

We are glad to share with you that we received lots of submissions from the conference and we selected a bunch of high-quality papers and compiled them into the proceedings after rigorously reviewing them. These papers feature but are not limited to the following topics: Remote Sensing, Global Positioning and Navigation System, Manned/Unmanned Aviation, High Performance Computing, Algorithms, etc. All the papers have been checked through rigorous review and processes to meet the requirements of publication.

We would like to acknowledge all of those who supported CRSA 2022 and made it a great success. Particularly, we would like to thank the Journal of Physics: Conference Series, for all the efforts of its colleagues in publishing this paper volume. We sincerely hope that the CRSA 2022 turned out to be a forum for excellent discussions that enable new ideas to come about, promoting collaborative research.

Committee of CRSA 2022

Committee member

Conference Chairs
Prof. Yanguo Jing, Coventry University, UK
Prof. Jiang Zhu, Tokyo Institute of Technology, Japan

Publication Chair
Prof. Vijayakumar Varadarajan, University of New South Wales, Australia

Organizing Committee Chairs
Dr. Chinthaka Premachandra, Shibaura Institute of Technology, Japan
Dr. Thippa Reddy Gadekallu, Vellore Institute of Technology, India

Organizing Committees
Prof. Lei Zhang, East China Normal University, China
Prof. Lishuai Jiang, Shandong University of Science and Technology, China
Prof. Quazi K. Hassan, University of Calgary, Canada
Prof. Shanzhong Qi, Shandong Normal University, China
Prof. Weifeng Liang, Guilin University of Technology, China
Prof. Wei Gao, Hohai University, China
Prof. Xianhong Chu, Qingdao Institute of Marine Geology, China
Prof. Xinfa Yang, Hainan Marine Geological Survey Bureau, China
Prof. Xingang Wang, Northwest University, China
Assoc. Prof. Bobo Li, Guizhou University, China

Technical Program Committees
Prof. Johan DEBAYLE, Ecole Nationale Supérieure des Mines de Saint-Etienne, France
Dr. Di Zhang, Henan University of Engineering, China
Dr. Deyu Yin, Huaiyin Institute Of Technology, China
Dr. He Xiao, Guilin University of Technology, China
Dr. Liqiang Wang, Institute of Mineral Resources, Chinese Academy of Geological Sciences, China
Dr. Lei Liu, Wuhan Center, China Geological Survey, China
Dr. Long Yuan, Petroleum Logging Co., Ltd. China
Dr. Xuliang Feng, Xi'an Shiyou University, China
Senior Engineer, Geological Exploration Company (Beijing), China
Senior Engineer, Li Liang, The Third Engineering Academy of Surveying and Mapping in Sichuan Province, China
Senior Engineer, Yanwen Chen, The first geological mineral exploration institute of Gansu provincial geology and mineral bureau, China

Journal of Physics: Conference Series **2410** (2022) 011002 doi:10.1088/1742-6596/2410/1/011002

Peer Review Statement

All papers published in this volume have been reviewed through processes administered by the Editors. Reviews were conducted by expert referees to the professional and scientific standards expected of a proceedings journal published by IOP Publishing Publishing.

- **Type of peer review:** Single Anonymous
- **Conference submission management system:** Morressier
- **Number of submissions received:** 47
- **Number of submissions sent for review:** 42
- **Number of submissions accepted:** 24
- **Acceptance Rate (Submissions Accepted / Submissions Received × 100):** 51.1
- **Average number of reviews per paper:** 2
- **Total number of reviewers involved:** 12
- **Contact person for queries:**
  Name: Xuexia Ye
  Email: xx.ye@keoaeic.org
  Affiliation: AEIC Academic Exchange Information Centre

Analysis and Research of Water Sprays in Civil Aircraft

Ming Liu

COMAC Shanghai Aircraft Design and Research Institute, Shanghai, China

e-mail: liuming5@comac.cc

Abstract—In the process of designing a new type of civil aircraft, the main manufacturer needs to obtain the type certificate, including the water spray test. Using the method of engineering estimation at the beginning of the design, we can effectively find the potential design problems in the initial stage and reduce the risk of subsequent tests. This paper studies water sprays under different runway water depths, tire parameters and speeds, obtains the basic track of the water sprays and determines the risk of the water sprays in compliance of civil aircrafts. It finds out the causes of the water sprays, analyzes them, and guides the model design.

1. Introduction

In rainy and snowy weather, the water and snow on the airport runway may cause the wheel water-skiing phenomenon when the aircraft is running. If the splash caused by the water-skiing is inhaled into the engine, it may cause the engine panting, stall, or even flame-out problems, and even lead to flight accidents[1][2][3]. On February 6, 1958, European Airways Flight 609 made a third failed attempt to take off from a flooded runway at Munich Airport in West Germany, killing 23 of the 44 passengers and crew on board. Since then, the condition of the airport runway has been valued since the accident.[1]The Netherlands National Aerospace Laboratory (NLR) in its 2001 report[2] The accident data of 136 airports in Western Europe is analyzed. There had been 91 flight accidents caused by runway pollution, including DC-10, MD-80, B727, B-737, and A320 aircraft. It is finally concluded that the accident rate caused by rain and snow pollution was more than 4 times higher than that of the normal runway. Flight accidents caused by runway pollution can be mainly divided into two categories, one is that the aircraft is out of control or even rushed out of the runway, and the other is the engine loses speed or even stalls caused by the water spraying into the engine. Wang Xiaomei studied the aircraft water sprays from the form, mechanism and influence.[3]Liu Chang and others have studied the aircraft water sprays by CFD.[4]Zhao Haigang and others have studied the aircraft water sprays from the point of test.[5]Xu Fei and others have studied the mechanism of water spray generated by the aircraft chine tire running on wet runway.[6]This paper mainly analyzes the situation of aircraft water sprays by using the method of engineering estimation in an early design.

Under AC20-124, the airworthiness provisions associated with the water sprays are CCAR25.1091(d)(2), and CCAR25.1325 (b), as described as follows:

CCAR25.1091 (d)(2)

For turbine engine-powered airplanes and airplanes incorporating auxiliary power units:

The airplane must be designed to prevent water or slush on the runway, taxiway, or other airport operating surfaces from being directed into the engine or auxiliary power unit air inlet ducts in hazardous quantities, and the air inlet ducts must be located or protected to minimize the ingestion of foreign matter during takeoff, landing, and taxiing.[7]

CCAR25.1325 (b)

Each static port must be designed and located in such a manner that the static pressure system performance is least affected by airflow variation, or by moisture, or other foreign matter. And that the correlation between the air pressure in the static pressure system and true ambient atmosphere static pressure is not changed when the airplane is exposed to the continuous and intermittent maximum icing conditions defined in Appendix C of this Part.[7]

2. Basic Consideration

According to the test conclusions of the previous aircraft, the actual spray area of the wheel is a three-dimensional space. Generally, the civil aircraft will have various functional atmospheric data sensors in the nose part, and the splash impact analysis should be carried out according to the airworthiness requirements. For a certain type of aircraft, its atmospheric data sensors include two total static pressure sensors, four static holes (two pairs), two total pressure sensors, two total temperature sensors, four angle sensors (two pairs), two full pressure probes, all of which are symmetrically mounted on both sides of the fuselage, as shown in Figure 1. No.1 static pressure hole, No.1.2 Angle sensor, No.1 full pressure probe, total pressure sensor, total temperature sensor are located in the nose part, far from the front, and the splash track should be the rear area of the wheel, so they will not be affected by the wheel splash. Just the impact of the splash on the No. 2 static pressure hole of the engine shall be considered. Only the impact of the splash on the No. 2 static pressure hole and the engine needs to be considered when conducting the analysis.

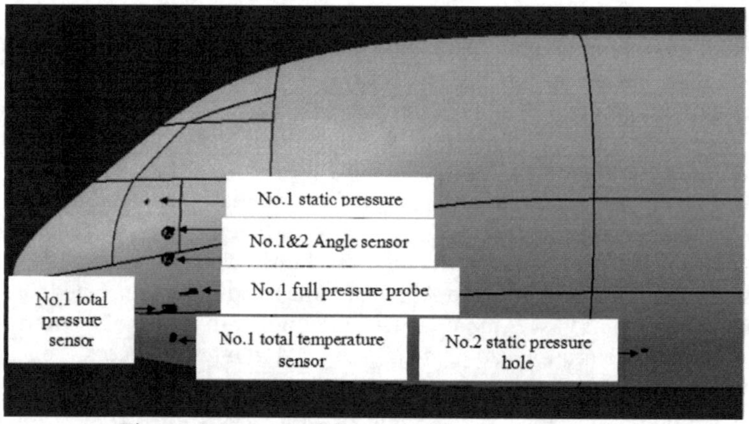

Figure.1 Layout of the atmospheric data sensor

When making the estimate, the area of impact is assessed by determining the water spray boundary.

The coordinate system used for water spray analysis moves with the aircraft. The origin of the coordinate system is located at the intersection between the projection line of the rotation axis of the plane and the symmetrical surface of the plane.

The water sprays boundary line is a spatial curve that can be decomposed into three planes of the water sprays coordinate system. If the track line projected from the outer boundary of the XY plane to the XY plane cannot reach the outlined line projected to the XY plane, it can be determined that the water sprays will not affect the engine.

Symbol definitions in this article are given in Table 1.

Table 1. Symbol definition

Symbol	Name	Unit
b_g	Width of the tire-ground contact area	inch
d	Depth of the fluid contaminant on the runway	inch
D	Diameter of an undeflected, inflated tire	inch
h	Half-length of the tire-ground contact area	inch
p	Tyre inflation pressure	psi
V_p	Sliding speed	knots
V	Ground speed	knots
w	Width of the undeflected, inflated tire section	inch
Z	Normal (to runway) load on the wheel	lbf
δ	Tyre vertical deflection	inch
θ_s	Side-spray plan view angle	deg
γ_s	Incremental spray angle used in the estimation of side-spray width	deg

3. Analysis Process

Water sprays track angle θs, γs calculation method of the water sprays point coordinates is shown in Figures.2 and 3.$F_{d_\theta} F_{\theta_s} \theta_{D_s} \gamma_{D_s} F_{\gamma_s} F_{d_\gamma}$ is the intermediate coefficient to calculate. X, Y is the water spray's point coordinate value. The specific formula is shown in Table 2.

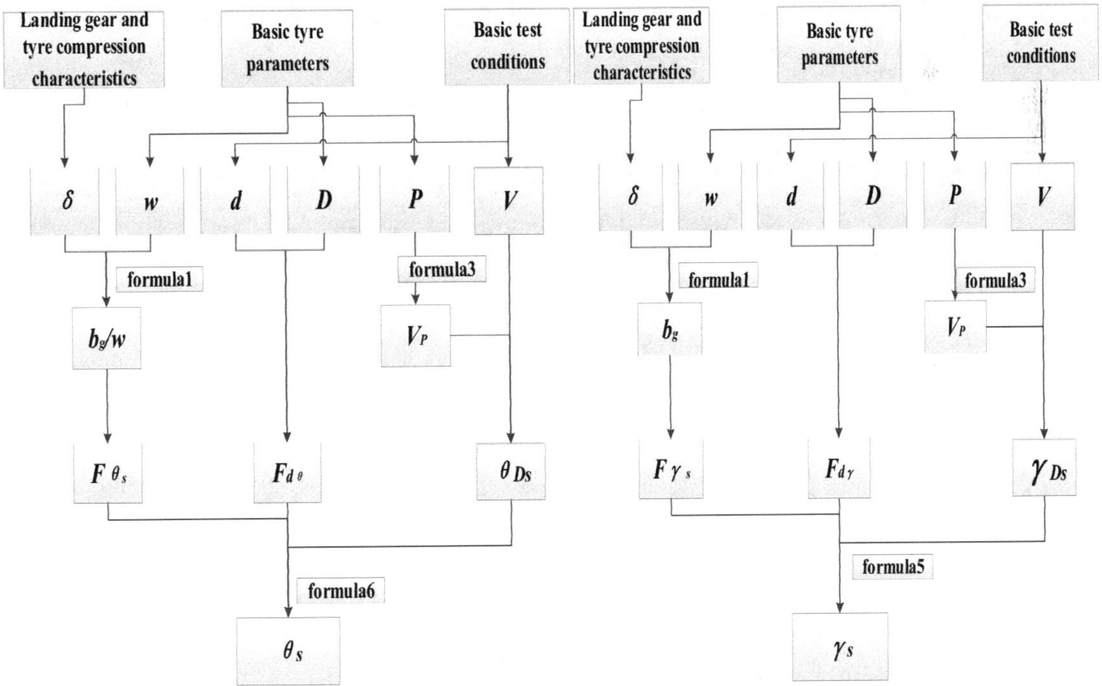

Figure.2 θs and γs computational process

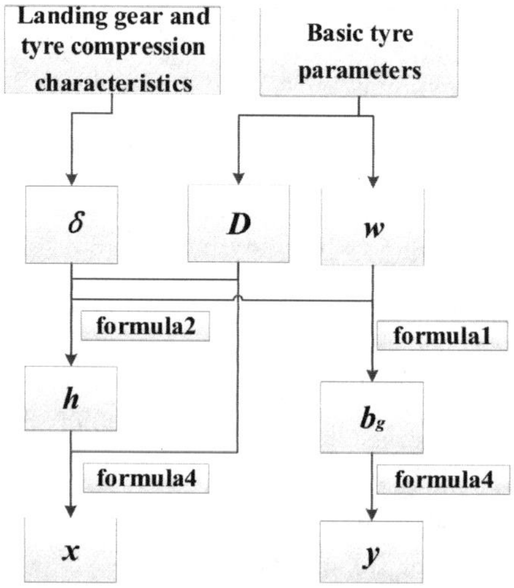

Figure.3 Water sprays point calculation process

4. Data Analysis

According to AC20-124: "Applicants seeking airworthiness approval can verify the water absorption characteristics of the engine and the air speed system with a test water area test of 1/2 inch of water depth."The runway water depths analyzed in this paper is based on 0.5in, and the comparative analysis is also conducted at 0.75 in and 1in water depths. At the same time, to explore the influence of the tire itself parameters on the water sprays trajectory, assuming that the compression characteristics do not change, based on the basic tire size, the front tire diameter D is 30 inches, width w is 8.8 inches, the front tire diameter D is 27 inches, width w is 7.75 inches, this parameter is the commonly used tire size of B737NG Series. In the analysis of this paper, takeoff weight is 78900 kg, front center of gravity 23%MAC. In the analysis, δ is 1.68 inches, and P is 178 psi.

4.1. Runway water depth is 0.5 inch

- The front tire has a diameter of 30 inches and a width of 8.8 inches.
  The calculation results are shown in Table 2.

TABLE 2. WATER SPRAYS ANALYSIS

V	60	70	80	90	100	110	120	130	140
θ_s	6.9	7.8	10.0	13.5	15.1	14.1	13.1	12.4	11.9
γ_s	11.1	12.4	14.2	15.8	15.5	14.0	12.7	11.7	11.0

- The front tire has 27-inch diameter and a width of 7.75 inches.
  The calculation results are shown in Table 3.

TABLE 3. WATER SPRAYS ANALYSIS

V	60	70	80	90	100	110	120	130	140
θ_s	7.3	8.2	10.5	14.2	16.0	14.9	13.9	13.2	12.6
γ_s	12.3	13.7	15.7	17.4	17.0	15.4	14.0	12.9	12.1

4.2. Runway water depth is 0.75 inches

- The front tire has a 30-inch diameter and a width of 8.8 inches.
  The calculation results are shown in Table 4.

TABLE 4. WATER SPRAYS ANALYSIS

V	60	70	80	90	100	110	120	130	140
θ_s	8.0	9.0	11.5	15.5	17.4	16.3	15.1	14.4	13.7
γ_s	13.9	15.5	17.8	19.7	19.3	17.5	15.8	14.6	13.7

- The front tire has a 27-inch diameter and a 7.75-inch width.

The calculation results are shown in Table 5.

TABLE 5. WATER SPRAYS ANALYSIS

V	60	70	80	90	100	110	120	130	140
θ_s	8.4	9.5	12.1	16.3	18.3	17.2	15.9	15.1	14.4
γ_s	15.2	16.9	19.4	21.5	21.1	19.1	17.3	15.9	15.0

4.3. Runway water depth is 1in

- The front tire has a 30-inch diameter and a width of 8.8 inches.
  The calculation results are shown in Table 6.

TABLE 6. WATER SPRAYS ANALYSIS

V	60	70	80	90	100	110	120	130	140
θ_s	8.7	9.8	12.6	17.0	19.1	17.9	16.6	15.7	15.0
γ_s	15.9	17.7	20.3	22.5	22.1	20.0	18.1	16.7	15.7

- The front tire has a 27-in diameter and a 7.75-inch width.

The calculation results are shown in Table 7.

TABLE 7. WATER SPRAYS ANALYSIS

V	60	70	80	90	100	110	120	130	140
θ_s	9.1	10.3	13.2	17.8	20.0	18.7	17.4	16.5	15.7
γ_s	17.1	19.1	21.9	24.3	23.8	21.6	19.5	18.0	16.9

4.4. Effects of the runway water depth

Using the basic tire parameters, the water spray Angle corresponding to different water depths is shown in Figure 4.

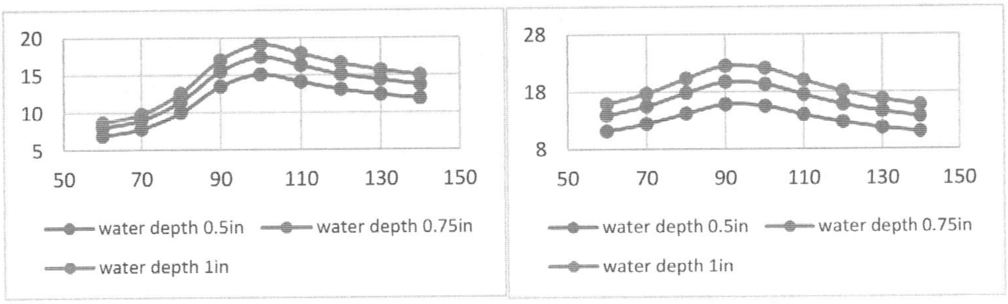

Figure.4 Effect of different water depths on θs and γs

Comparative analysis shows that both θs and γs become larger as the water depth increases.

4.5. Effects of the tire parameters

The water depth was analyzed at 0.5 inches, and the water sprays angles corresponding to the standard and contrast tires are shown in Figure 5.

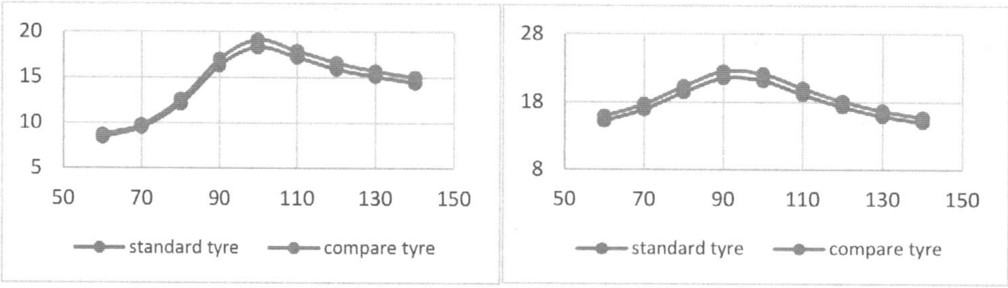

Figure 5. Effect of different tires on θs and γs

Through comparative analysis, it can be found that compared with standard tires, both θs and γs are produced in comparative tires.

From the above calculation, it can be found that under different runway water depths, and tire parameters and θs Maximum value are between 15.1° to 20.0°, and γs Maximum value is set at 15.8° to 24.3°, as shown in Figure 6.

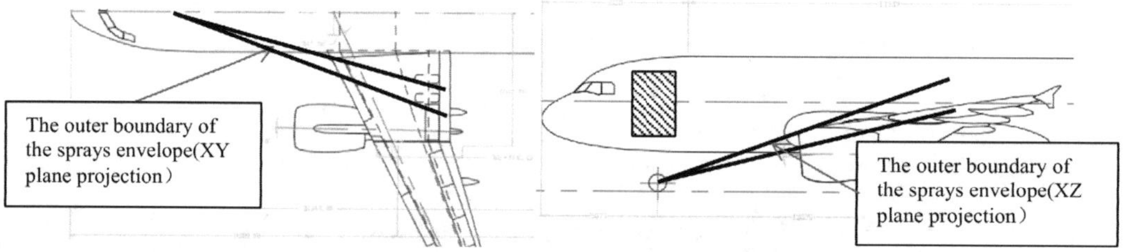

Figure 6. Water sprays boundary

5. Conclusion

In this paper through the analysis of the civil aircraft water sprays under different runway water depths and tire parameters, more reliable water sprays curve data is obtained, which clarifies the impact of runway water depth and tire parameters on water sprays, which has a certain reference value for model design and test:

Generally speaking, the aircraft with the engine under the wing water sprays test risk is small, the engine lip margin from the outer boundary of the water sprays package line, for a certain model, even in the most severe conditions, there are still 10° Left and right remainder.

The water depth has a great impact on the water sprays Angle, and the water sprays range increases with the water depth. A model is taken as an example, the water depth increases by 0.25 inches, and the maximum side angle and overlooking angle increase by 2° on the left and right sides.

Tire parameters affect the water sprays Angle, and the water sprays range increases with the decrease of tire size. Taking a certain aircraft as an example, the wheel size is reduced from 30*8.8 to 27*7.75, and for different water depths, the maximum side perspective and top viewing Angle increase by 1 to 4.

References

[1] Van Es G W H, Roelen A L C, Kruijsen E A C, et al.Safety aspects of aircraft performance on we t contaminated runways[C].10th Annual European Aviation Safety Seminar.Amsterdam: Na tional Aerospace Laboratory,1998
[2] Dreber C R,Horne B W.Ground-run tests with a bogie landing gear in water and slush,NASA TN D-3515[R].Washington DC,1966
[3] Qi X F, Zeng T.Water ingestion certification test method of civil aircraft power system[J].

[4] Zhang Y, Xu Fi, Duan M, Gao X, Lu J, Ding W. Calculation method of aircraft tire water sprays and water retaining principle analysis of turning edge tires [J]. Science, Technology and Engi neering, 2014,2 (6): 54-59
[5] Van Es G.W.H, Roelen A.L.C, Kruijsen E.A.C, et al.Safety aspects of aircraft performance on we t and contaminated runways.NLR Technical Paper 216. 2001
[6] Wang X M, Aircraft tire splashing mechanism[J].Technology Innovation and Application, 2018(2 6): 49-50
[7] Liu C, Chen Y Y, Shuai C. Simulation Analysis of Aircraft Tire Splash and Application[J].MEC HANICAL ENGINEERING&AUTOMATION, 2019,12(6): 112-115
[8] Zhao H G, Qu J Y, Ma S, Fu X G, Qi X F. Water ingestion test technical of transport aircraft pow er system[J].Journal of Aerospace Power, 2021,12(12): 2674-2682
[9] Xu F, Ren X Q, Li Y N, Gao X Y, Lv J.Mechanism of Water Spray Generated by Aircraft Chine Tire Running on Wet Runway[J].Journal of Northwestern Polytechnical University, 2017,8 (4): 616-621
[10] CIVIL AVIATION REGULATION OF CHINA. PART 25 AIRWORTHINESS STANDARDS: TRANSPORT CATEGORY AIRPLANES[S].CIVIL AVIATION REGULATION OF CHI NA, 2011

Research on Testing and Calculation Analysis of Cargo Bay Fire-Extinguishing Agent Concentration Based on Different Loading in Civil Aircraft

Hongfei Kou

China COMAC Shanghai Aircraft Design and Research Institute, Shanghai 201210, China

kouhongfei@comac.cc

Abstract—In the history of international aviation, there have been many fire accidents of the cargo compartment. To ensure that the fire is extinguished in time when the cargo compartment is on fire, and to ensure the safety of aircraft and personnel, the design of the cargo compartment fire extinguishing system needs to be verified according to CCAR 25. According to the AC851-1 guidance, the verification of the cargo compartment's fire extinguishing system design aims to test and calculate the concentration of fire extinguishing agents used for the C-class cargo compartment based on different loading conditions of civil aircraft. Through the flight test of the cargo compartment under a no-load condition, it can be seen that the concentration of fire extinguishing agents in the cargo compartment is higher than 5% in the initial phase, higher than 3% in the 39,000ft and 10,000ft level flight phase, and the entire flight test phase is maintained above the suppression concentration required by the test, which meets the system design requirements. The concentration of fire extinguishing agents in the cargo compartment under a no-load condition is calculated and analyzed according to the maximum load rate (full load), and the result is higher than the required 3% suppression concentration.

1. Introduction

During the operation of civil aircraft, a large amount of passenger luggage and cargo is stored in the C-class cargo compartment. Therefore, there have been many cargo compartment fire accidents in the history of international aviation operations. According to statistics, the cargo compartment fire probability is one in a million per flight. In the event of a fire in the C-class cargo compartment, not only is the loaded cargo damaged, but also flight safety is even endangered, which directly leads to the crash of the aircraft. When the cargo compartment catches fire, the high-speed fire extinguishing bottle is usually released first to put out the fire. Since the inside of the bottle is a flammable material to avoid re-ignition before the aircraft lands, the low-speed fire extinguishing bottle should be released at the same time to continuously replenish fire extinguishing agents in the cargo compartment, and to provide effective fire suppression means for the cargo compartment until the aircraft lands safely.

Usually, the airworthiness certification of civil aircraft in my country referrs to obtain the CAAC (Civil Aviation Administration of China) type certificate mainly through the CCAR25 airworthiness regulations, and the airworthiness verification of the C cargo compartment fire extinguishing system of civil aircraft is an important part of the civil aircraft certification work. According to §25.851(b)(2), §25.855(h)(2), and other provisions of CCAR25 and AC851-1, evaluation and related computational analysisis are carried out to find out whether the fire extinguishing system of the cargo compartment meets the expected function and safety airworthiness requirements, mainly through the concentration test of fire extinguishing agents in the cargo compartment. The FAA (Federal Aviation Administration, Federal Aviation Administration) issued AC851-1 in 2016 and given the guidance and recommendations. For the test of the concentration of fire extinguishing agents in the cargo compartment, the influence of different loading conditions on the concentration of fire extinguishing agents should be considered. The concentration of fire extinguishing agents under a No-load condition is verified by the flight test method, and other loading conditions need to be analyzed using its recommended calculation method to show that the concentration of fire extinguishing agents in the cargo compartment meets the airworthiness requirements [1]. Therefore, based on a certain type of project, the concentration of fire extinguishing agents (Halon1301) in the C-class cargo compartment of the civil aircraft under different loading conditions is tested and calculated, as well as analyzed to verify that the design of the C-class cargo compartment fire extinguishing system for this type of project meets the requirements of relevant airworthiness regulations, and at the same time to provide support and foundation for the follow-up airworthiness technology of domestic civil aircraft cargo compartment's fire extinguishing system.

2. No-load condition flight test

According to the guidance and recommendations in the AC851-1, the C-class cargo compartment's fire extinguishing agent concentration flight test should be carried out under the condition of an empty cargo compartment, and under other loading conditions to verify that the cargo compartment fire extinguishing agent concentration still meets the airworthiness requirements through calculation and analysis methods to ensure flight safety. According to the requirements of airworthiness regulations, using the Halon1301 cargo compartment fire extinguishing system, the flight test pass criterion is that the initial volumetric concentration of fire extinguishing agents in the cargo compartment is not less than 5%, and the volumetric concentration of fire extinguishing agents in the subsequent cargo compartment must be kept at least 3% to suppress the fire in the cargo compartment until the plane lands safely.

2.1. Flight condition

In the test, the maximum cabin pressure difference was selected as the most severe test condition. In the test, the loss of fire extinguishing agents mainly included two aspects: 1) fire extinguishing agents in the cargo compartment leak to the outside of the fuselage; 2) fire extinguishing agents leak to the triangle area of the cargo compartment. Since the cargo compartment and the cargo compartment triangle area belong to the pressurized area, the pressure difference remains the same under various pressurized states, and the loss of fire extinguishing agents to the cargo compartment triangle area is less; the pressure difference between the cargo compartment and the outside of the aircraft is large, fire extinguishing agents were mainly lost to the outside of the fuselage.

The selection of the test flight altitude is very important. The cargo compartment fire extinguishing agent concentration test needs to verify the leakage of the cargo compartment's fire extinguishing agent concentration under a condition of the maximum pressure difference. Therefore, for the flight test of this project an altitude of 39000ft (the maximum cruise altitude) was selected, and the dual air conditioners worked. After the fire extinguishing bottle in the cargo compartment was released and flew for a while at 39,000ft, the aircraft descended to an altitude of 10,000ft until the end of the test.

According to the AC851-1, the flight time (approved protection time) of the cargo compartment's fire extinguishing agent concentration test should be greater than the design yaw time: ETOPS (Extended-Range-Operation) time + 15min (approach and landing time), the ETOPS time is 60 minutes for this type of the aircraft, and 5 minutes of additional time is considered, so the duration of the entire test flight is 80 minutes. According to the descent rate and standard approach time of this type of the aircraft, the duration of the aircraft's descent time, approach and landing time is about 20 minutes. Therefore, in this flight test, the flight time at 39000ft is selected as (75min-20min) for a total of 55min, and the flight time at 10000ft is 25min (with 5min as the margin).

The leakage rate of the cargo compartment has a great influence on the test results of the concentration of fire extinguishing agents in the cargo compartment. According to the AC851-1, it is necessary to consider the influence of the aging of the cargo compartment seal on the leakage rate. The sealing material and fabric composition used in this aircraft have been verified to meet the aging requirements during the service period. In conclusion, it can be judged that the materials used have no effect on the air-tightness during the entire service period of the aircraft, and will not affect the leakage of fire extinguishing agents in the cargo compartment, so new seals are selected in the test.

2.2. Test equipment

Extinguishing agent concentration test equipment is the key to the collection of fire extinguishing agents in the flight test. The Statham GA.2A gas concentration analyzer is used in this test, which has been approved by the FAA. The test analyzer has the characteristics of wide application and high sampling accuracy. Other auxiliary test equipment includes a recording oscilloscope, vacuum pump, controller, sampling probe and connecting cable. Its working principle is that the vacuum pump draws the gas sample to the analyzer through the sampling probes distributed in the cargo compartment. Each analyzer is composed of 4 analysis chambers. The analyzer is used to measure the pressure drop of the porous metal plug inside the device. The pressure drop is a function of gas viscosity and flow rate, and the sensor components in each analysis chamber generate a signal as a function of pressure drop, which is sent to the processor (oscilloscope) [2-5]. The test equipment connection is shown in Figure 1.

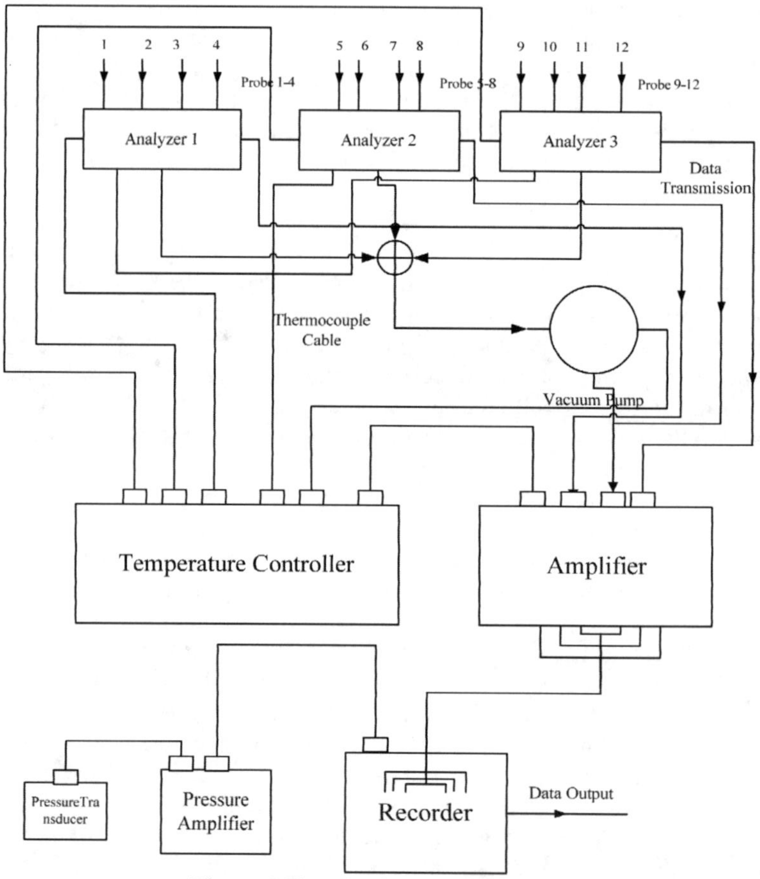

Figure 1 Test Equipment Connection

2.3. *Test installation*

The concentration of fire extinguishing agents in the cargo compartment is obtained by analyzing fire extinguishing agents extracted from their test sampling probe installed in the cargo compartment. The Statham GA.2A gas concentration analyzer uses 12 test sampling probes, so the selection of the sampling area must be carefully considered, with the correct installation and configuration of the test sampling probe performed to ensure that fire extinguishing agents' concentration test data is representative.

Before the test, 11 test sampling probes, 1 return pipe, 1 pressure sensor and 1 thermocouple were installed in the cargo compartment, and 1 test sampling probe was installed in the passenger cabin. The purpose of installing the sampling probe in the passenger cabin was to confirm whether the concentration of fire extinguishing agents in the cabin exceeded the standard. Normally, the concentration of extinguishing agents that should be spread to the passenger cabin shall not exceed 5% by volume. Three test sampling tube stanchions are installed in the cargo compartment, and the sampling probes are fixed on the sampling tube stanchions by adhesive tape.

2.3.1. *Placement of stanchions*

The placement of the test sampling probe stanchions in the cargo compartment is as follows. The installation position is shown in Figure 2.

(1) The first sampling probe stanchion is arranged at a position of 1900mm away from the front plate of the cargo compartment and near the centerline of the cargo compartment;

(2) The second sampling probe stanchion is arranged at a position of 3800mm away from the front plate l of the cargo compartment and near the centerline of the cargo compartment;

(3) The third sampling probe stanchion is arranged at a position of 1900mm away from the rear plate of the cargo compartment and near the centerline of the cargo compartment.

2.3.2. *The placement of sampling probes*

The location of the test sampling probe is as follows. The installation position is shown in Figure 2.

(1) In the first set of test sampling probes (Probes 1-4), four test sampling probes are vertically arranged on the first stanchion, with heights of 88mm, 315mm, 620mm and 150mm from the cargo roof;

(2) In the second set of test sampling probes (Probes 5-8), four test sampling probes are vertically arranged on the second stanchion, with heights of 88mm, 315mm, 620mm and 150mm from the cargo roof;

(3) In the third group of test sampling probes (Probes 9-11), the three test sampling probes are vertically arranged on the third stanchion, and the heights are 88mm, 315mm and 150mm from the cargo roof;

(4) The sampling probe (Probes 12) of the passenger cabin is located on the right side of the corresponding middle position just above the cargo compartment in the passenger cabin.

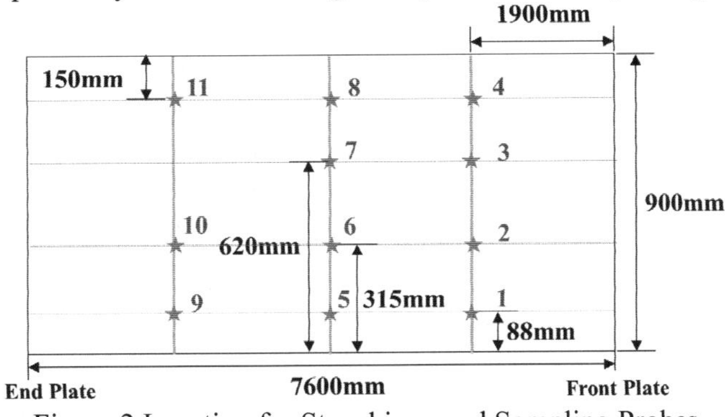

Figure 2 Location for Stanchions and Sampling Probes

2.4. *Test result and analysis*

Figure 3 shows the results of the flight test under a no-load condition in the cargo compartment. analysis as below:

（1）The initial fire extinguishing agent concentration of the test sampling probes (Probe 1~Probe 11) in the cargo compartment is always higher than 5%, and the lowest value is 5.7%, which appears in Probe 4;

（2）During the 39000ft and 10000ft level flight phase, the concentration of fire extinguishing agent in the test sampling probes (Probe 1~Probe 11) in the cargo compartment is always higher than 3%, and the lowest value is 7.4%, which appears in the test sampling Probe 1;

（3）The concentration of fire extinguishing agents in the test sampling probe (Probe 12) in the cabin of the 39000ft and 10000ft level flight phase is always less than 5%, and the highest value is 1.7%.

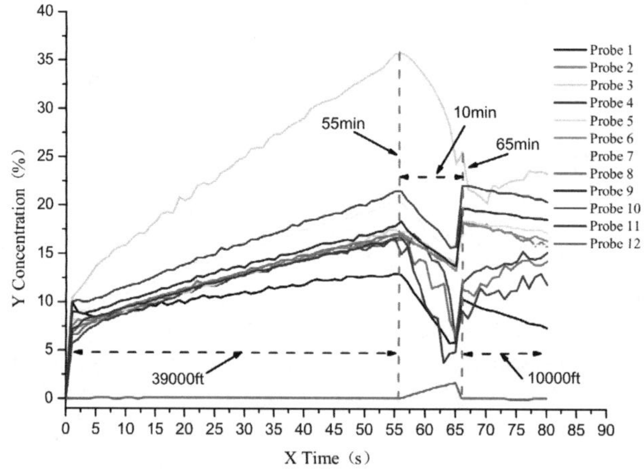

Figure 3 Test Results under a No-load Condition Flight Test

3. Calculation and Analysis of Load Condition

According to the recommendations in the AC851-1, the concentration of fire extinguishing agents in the cargo compartment under non-empty conditions should be determined by calculation and analysis methods. As the loading rate of the cargo compartment increases, the volumetric ratio of the cargo compartment decreases, which leads to an increase in the initial concentration of fire extinguishing agents in the cargo compartment. Since the leakage rate remains unchanged, the concentration of fire extinguishing agents in the cargo compartment will decrease faster as the loading rate increases. Generally, the higher the loading rate is, the faster the concentration descends. In this paper, the maximum loading rate (full load) is selected as 90% for calculation. The formula is as formula (1):

$$C(t)=\left[C(0)-\{RS/E\}\right]e^{-Et/V}+\{RS/E\} \quad (1)$$

In the formula, relevant parameter definitions and values are shown in Table 1.

Table 1 Parameter Definitions and Values

Parameter	Description	Unit	Value
C(t)	Halon 1301 extinguishing agent concentration at t minutes after agent release	N/A	
C(0)	Halon 1301 extinguishing agent concentration at the end	N/A	0.42

	of Phase I		
R	Average release flow rate of orifice Halon 1301 extinguishing agent	lbs/min	0.5
S	Halon 1301 extinguishing agent specific volume	ft^3/lbs	2.36
E	Cargo compartment leakage rate	ft^3/min	25
V	Cargo compartment left volume	ft^3	51.71
C(0)-RS/E		N/A	0.38
RS/E		N/A	0.046

Calculation and analysis are conducted according to the formula (1), and the result is shown in Figure 4.

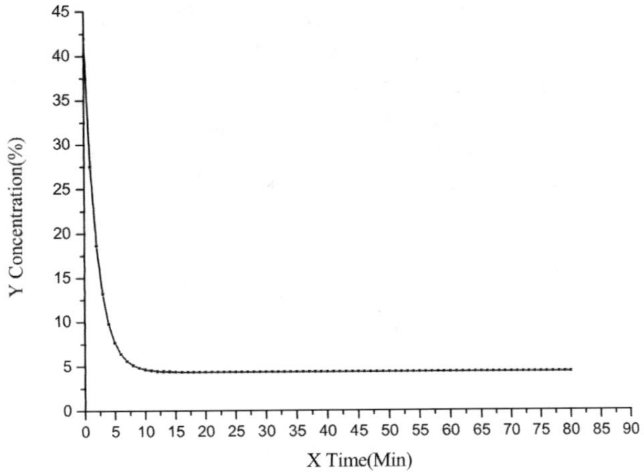

Fig.4 Analysis Results of the Concentration of Fire-Extinguishing Agents when the Cargo Compartment is fully Loaded

From the results in Figure 5, it can be seen that the concentration of fire extinguishing agents in the cargo compartment is very high at the initial stage of the release of agents, and because the leakage rate of the cargo compartment remains unchanged, the subsequent concentration of fire extinguishing agents decreases very quickly, which is completely consistent with the description of the AC851-1. With the passage of time, when fire extinguishing agents were released for more than ten minutes, the concentration of fire extinguishing agents in the cargo compartment reached equilibrium and remained in the range of 4.3% volumetric concentration until the end of the calculation. Therefore, based on the above analysis, it can be seen that the design of fire extinguishing agents concentration in the cargo compartment of this aircraft type meets the requirements. Even in the case of the maximum loading rate, the cargo compartment's fire extinguishing agent concentration still meets the requirement of at least 3% volumetric concentration during the entire flight phase.

4. Conclusion

According to the guidance of the AC851-1, the concentration of fire extinguishing agents in the C-class cargo compartment of civil aircraft based on different loading conditions is tested, calculated and analyzed. Through the flight test of the empty cargo compartment, it can be seen that the concentration of fire extinguishing agents in the cargo compartment of a certain type of aircraft remains higher than the concentration required by the test during the entire flight test stage, which meets the system design requirements. In the test, the test modification method, test condition selection and test results analysis method are established. The concentration of fire extinguishing agents in the cargo compartment under non-empty conditions is calculated and analyzed based on the maximum loading rate (full load), and the result is higher than the required suppression concentration.

This project has an important guiding role in the test verification,calculation and analysis technology of fire extinguishing agent concentration in the cargo compartment of the civil aircraft and provides technical support for the airworthiness and certification of the civil aircraft cargo compartment's fire extinguishing system. Follow-up in-depth research will be carried out on the test method and fire extinguishing nozzle design, with a view of improving the concentration of fire extinguishing agents by optimizing the design of the cargo compartment's fire extinguishing system and ensuring that civil aircraft can fly more safely.

References

[1] AC25.851-1 Build-in Fire Extinguishing/Suppression Systems in Class C and Class F Cargo Compartments,2016.
[2] AC 20-100.General Guidelines for Measuring Fire-Extinguishing Agent Concentrations in Power plant Compartments [S]. Department of Transportation Federal Aviation Administration, 197 7-09-21.
[3] Lee s, Frick K, Lamb M. In-flight engine fire-extinguishing agent concentration tests for commercial airplane [R].AIAA-95. 3977. 1995.
[4] Criteria for aircraft installation and utilization of an extinguishing agent concentration reorder[R].FAA Final Report No.FAA-DS-70-3.(AD-712-19).1970.
[5] Qian Zhonghe.The simple verification method for aircraft fire-extinguishing system performance[J].Civil aircraft design and research,1993.

Function Design of Civil Aircraft Brake System Based on Model

Renyu Huang[1a*], Haomin Li[1b*]

[1]COMAC Shanghai Aircraft Design and Research Institute, Shanghai 201200, China

[a*]964770682@qq.com; [b*]LIHAOMIN@comac.cc

Abstract— In view of the practical problems in aviation field, such as increasingly complex systems and relationships among systems, great uncertainty in cross-industry and multi-discipline, traditional systems engineering is gradually weak. In this paper, combined with the functional architecture design of civil aircraft brake system, model-based systems engineering (MBSE) is introduced. In this paper, the function of civil aircraft brake system is defined and analysed by this method, and the functional architecture model of civil aircraft brake system is built through repeated iterations, and the state switching logic among system functions is presented digitally to ensure the clarity, comprehensiveness and reliability of the functional architecture.

1. Introduction

Engineering [1-2] is an interdisciplinary method that integrates development, manufacturing, operation and management from top to bottom. It simulates the iterative process of real systems, and achieves global optimization or approaches global optimization to meet the system requirements. However, the traditional document-based system engineering has inherent defects in describing the system architecture model: the integrity and consistency of information are difficult to evaluate and determine; It is difficult to describe various activities; It is difficult to change; The iteration period is long.

To solve the above problems, a new method (model-based system engineering) came into being. MBSE [3-4] is a method to realize the relationship between systems by modeling. The key of MBSE is to establish a systematic and standardized model, in which we should apply the model paradigm in the whole process, establish a unified database, and correlate parameters online. Compared with traditional system engineering, MBSE has many advantages, such as ambiguity of knowledge representation, reusability of system content and breaking the time and space constraints. In this paper, MBSE is applied to the brake system, focusing on the functional design, defining the functions from the aircraft level to the system level, and building a functional architecture model. Based on the functional architecture, a functional mechanism model is built and verified by simulation.

2. Brake system

The Brake system [5] is composed of a green circuit, yellow circuit, brake temperature monitoring system and tire pressure monitoring system. The brake hydraulic sources 1 and 2 provide hydraulic energy for the green circuit and the yellow circuit respectively. The brake device is in the form of a semi-double cavity, which is connected with the green and yellow circuits respectively. The green circuit and the yellow circuit are physically independent and completely isolated from each other; The two circuits work simultaneously. The green circuit includes a brake accumulator, cut-off valve, brake control valve and other components. The yellow circuit includes a brake accumulator, shut-off valve, parking brake valve, normal/parking brake control valve and other components. Two circuits are equipped with pressure accumulators; Ensure that it still has partial braking capacity when a set of hydraulic energy is lost. The two brake circuits are controlled by two BCU respectively, and the pilot can brake by stepping on the brake pedal or selecting the corresponding automatic brake gear.

A civil aircraft brake system is a complex system that integrates machinery and control, and as an important part of the aircraft, it is necessary to ensure the safety and reliability of the system, so as to provide safe braking for the aircraft in various situations. Because of the high complexity of the brake system, MBSE method is needed to improve the completeness and correctness of the system architecture and reduce the risk in the research and development process.

3. Functional architecture

3.1. Systems and functions

When we describe the braking system, we should base it on two different perspectives. One is to describe the system from the internal physical structure of the braking system, which requires the braking subsystem or physical equipment to complete the system tasks under different environments and conditions; The other is to describe the system from functions. The braking system needs to meet the expected functions of stakeholders.

The word "function" is no stranger to most people, and it is an attribute that an object can meet a certain demand. It is often described in the phrase form of "verb+noun", for example, "provide control of unit operation and data input" and "support ground operation". Function is not an isolated concept, and its meanings are different in different fields and levels. For the aircraft system, or the civil aircraft brake system, the functions need to include at least the following three aspects:

(1) Effect of function: define the requirements corresponding to the function, and know the goal to be accomplished by the function;

(2) Status of functions: Understand the status switching of functions, and distinguish the functions that are easy to describe from those that are relatively difficult to describe;

(3) Realization of function: master the mechanism of function realization. This step is also an in-depth study of the system mechanism.

3.2. Definition of functions

Modern civil aircraft generally adopt the top-down R&D process. Starting from the determination of requirements [6], it carries out function definition, aircraft scheme design, system design, component design and production, and then it is integrated, tested, tested and verified, and finally delivered to users. This can ensure the integrity of the design and the compatibility between the various systems, which will bring benefits to the subsequent integration, verification and airworthiness forensics, and also facilitate us to strengthen our understanding of the whole aircraft system and individual functions.

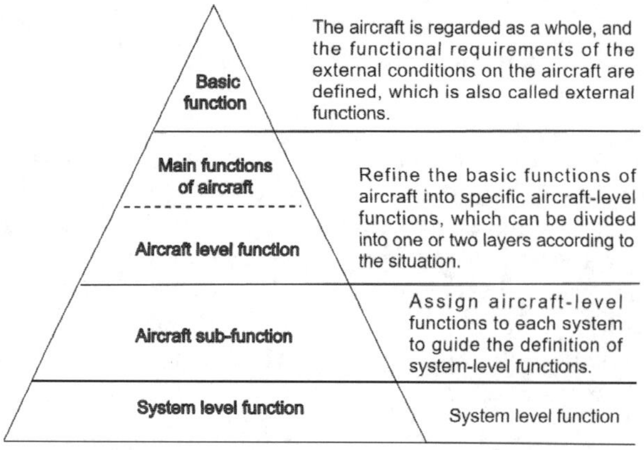

Figure 1 Functional hierarchy allocation

The definition of aircraft function [7-8] plays a connecting role in this R&D process. The quality of function definition directly determines whether the function model can systematically describe the whole system. It is related to concepts such as functional objectives and behaviors, and defines functions as well as these correlations. The definition of functions should not only pay attention to the characteristics of different systems, but also aim at solving problems according to requirements, so that the definition of requirements will not be too narrow or too broad. For civil aircraft, the definition of functions mainly focuses on aircraft-level functions and system-level functions. The relationship of several levels is shown in Figure 1.

Aircraft level functions are subdivided into four levels: basic function, main function, function and sub-function. The first basic function is to regard the aircraft as a whole, from the external perspective, to ensure the normal operation of the aircraft in any scene, and this layer is also the basis of aircraft functional modeling; The second and third layers are to refine the external functions of the aircraft into specific aircraft-level functions, mainly focusing on the related operations of the people on the aircraft. In different environments, the pilot's operations are also different. In the process of the aircraft moving from the starting point to the end point, the pilot needs to control the aircraft. When on the ground, the pilot needs to control the aircraft on the ground. The fourth layer is to assign aircraft-level functions to each system to guide the definition of system-level functions.

In different flight stages and different environments, the functions used by aircraft are different. This study defines the functions from the scene related to the braking system, maps the functions from the aircraft level to the braking system, and finally obtains three related sub-functions of the aircraft level: ground deceleration; providing group operation control and data input; supporting ground operation.

Table 1 Aircraft level function

Basic Function	**Main Function**	**Function**	**Sub-function**
Moving from the starting point to the finishing point	Control aircraft	Ground control aircraft	Ground deceleration
Provide public resources on the plane	Provide man-machine interface	Provide interface for flight crew.	Provide unit operation control and data input.
Support, maintenance and route improvement	Support aircraft route operation	Support ground operation	Support ground operation
...	...	...	...

Table 1 is the basic decomposition result of the aircraft-level function of the braking system. This result is extremely important for the whole aircraft, but it is a bit simple for the aircraft architecture design and needs further improvement. According to the airplane sub-function, combined with the design objectives and requirements of wheel braking system, the braking system-level functions are decomposed, and the system and functions are defined in detail. For example, the sub-functions of ground deceleration can be further divided into six system-level functions: pedal braking, automatic braking, anti-skid function, braking monitoring, absorbing braking energy and providing braking system information. See Table 2 for details.

Table 2 System level function

Aircraft level function	System level function	Functional description
Ground deceleration	Pedal brake	The main brake should provide the pilot with the ability to stop the plane with the brake pedal.
	self-actuating brake	Automatic braking should provide the pilot with the ability to stop the plane without using the brake pedal when on the ground.
	Anti-skid function	Brakes should have anti-skid function.
	Brake monitoring	The wheel brake system shall monitor the brake temperature and tire pressure of each wheel.
	Absorb braking energy	The wheel brake system should have the ability to absorb braking energy.
	Provide brake system information.	The wheel brake system should provide information needed by other aircraft systems to support the operation of these systems.
Provide unit operation and data input.	Provide man-machine interface for maintenance personnel.	Provide brake system status and information to maintenance personnel.
Support ground operation	Provide ground traction.	The brake system shall provide ground traction capacity.
	Stop braking	Parking should be able to provide enough braking to prevent the aircraft from sliding on the runway.

3.3. *Functional architecture model*

The functional architecture model is not the random accumulation of knowledge, but the organized and logical organization of knowledge in some way. Like other models, the functional architecture model is also a graphical representation, whose purpose is to describe functions and the information interaction between functions. This paper mainly describes the functional architecture of the brake system in the form of BDD (Block Definition Diagram) and IBD (Internal Block Diagram).

BDD is the most common SYSML (System Engineering Modeling Language) [9] diagram. Different types of model elements and relationships can be displayed in BDD to express system information. According to the functions included in the brake system and its traced airplane-level functions, the BDD of the brake system is shown in Figure 2.

IBD and BDD are closely related. In IBD, elements can be displayed to describe all aspects of the system, which can be regarded as a supplement to BDD. Besides, IBD has its own unique functions. The IBD of the architecture mainly describes the functional interface requirements captured in the work activity diagram in the operation scenario, and presents the logic between functions in the form of interfaces and interactive items, so as to describe the information flow trend between functions and external roles. The braking system ground deceleration function IBD is shown in Figure 3.

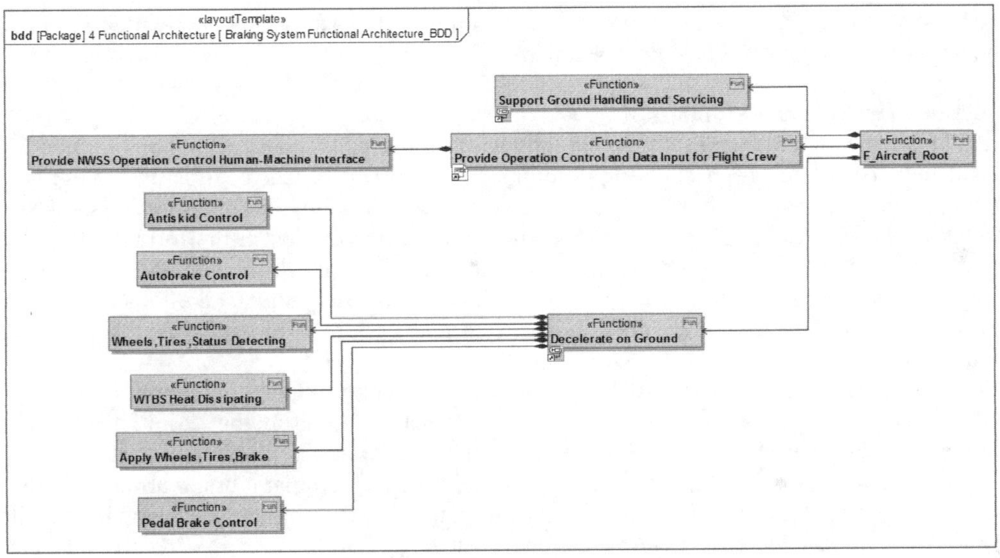

Figure 2 Braking system BDD

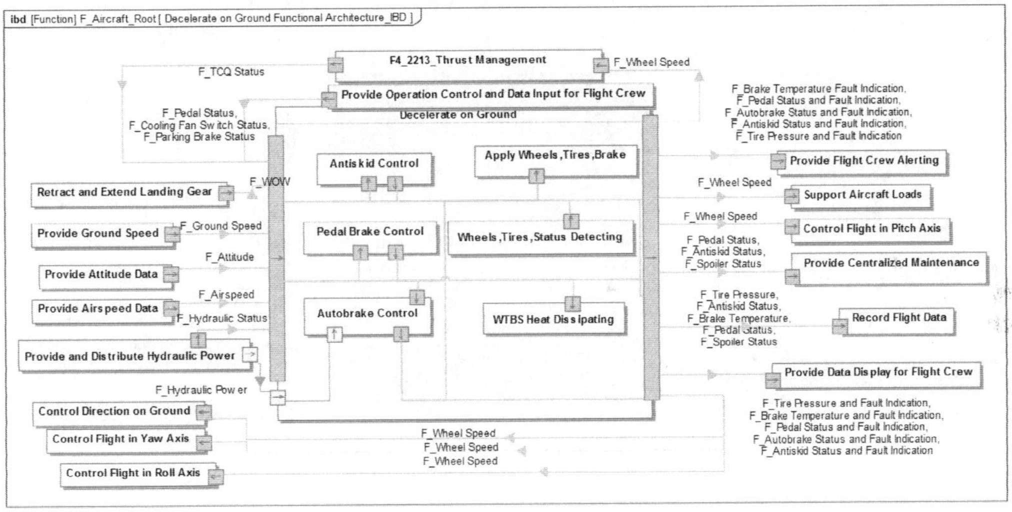

Figure 3 Braking system IBD

4. Logical definition

The logic definition mainly describes functions in three dimensions: activity, sequence and state. Modeling first describes the three functions of ground deceleration, providing unit operation and data input, and supporting ground operation from a black-box perspective, and uses state machine diagram (STM) to express the state switching and switching logic among functions. STM generally consists of several sub-states that represent system operation, and defines a sub-state as the starting position. At the same time, according to the set conditions, the functions are switched from one to the other. On the one hand, STM shows the normal state switching, on the other hand, it also shows the influence of external roles on the internal state of the system.

The last stage of modeling is activity diagram (ACT) modeling, which relies on state machine modeling, describes the activities of each state, and assigns the requirements to the architecture, thus completing the transformation of brake system functional architecture from black box perspective to white box perspective. Activity diagram describes the workflow of the system at runtime, and defines the logic judgment generated in the workflow. The output of this stage pays more attention to details, and the description of the system is more detailed.

5. Functional architecture simulation

After the brake system functional architecture model is established, its dynamic behavior, running logic, input and output need to be checked and simulated to confirm the correctness and integrity of the architecture model and ensure the integration with subsequent systems. In this paper, MagicDraw [10] is selected as the simulation software, and the simulation is carried out in the form of SYSML state diagram, to verify the correctness of its logic and data transmission. In the simulation experiment of brake system functional architecture, the designer needs to manually simulate the joint simulation with other systems, and judge the working state and logic correctness of the model by inputting defined data or events.

Take the simulation of STM and ACT, which provide pedal braking and automatic braking, as an example. When simulating automatic braking, the system starts to run from the set initial position (brake off). When receiving wheel load signal and pedal signal, the system will enter the automatic braking module. The automatic brake module starts to operate from the automatic brake arming, as shown in Figure 4. In the arming activity diagram, when the arming is successful, the automatic brake will send out the signal of throttle lever to the idle position. After the system receives it, it receives the signal of wheel speed of 120 knots, and the system enters the automatic brake opening mode, and the automatic brake starts to work, as shown in Figure 6. During this process, the changes of key values of the system are shown in Figure 5.

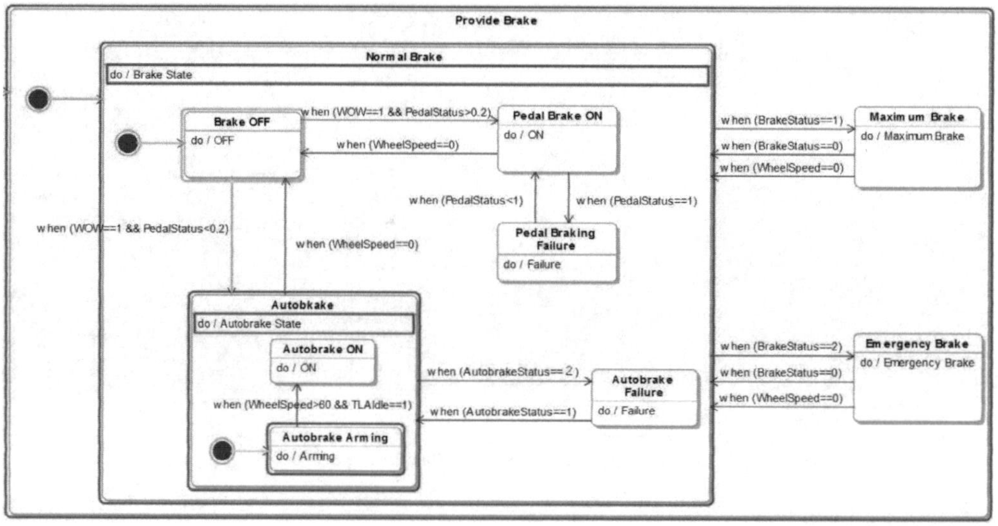

Figure 4 STM

Figure 5 Key numerical variation

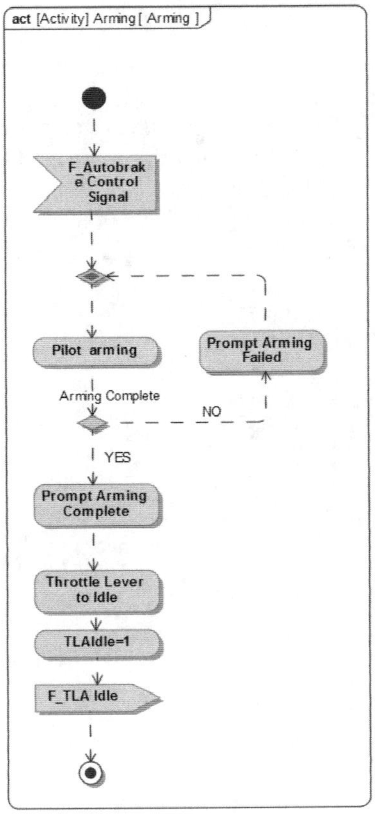

(a) Autobrake arming

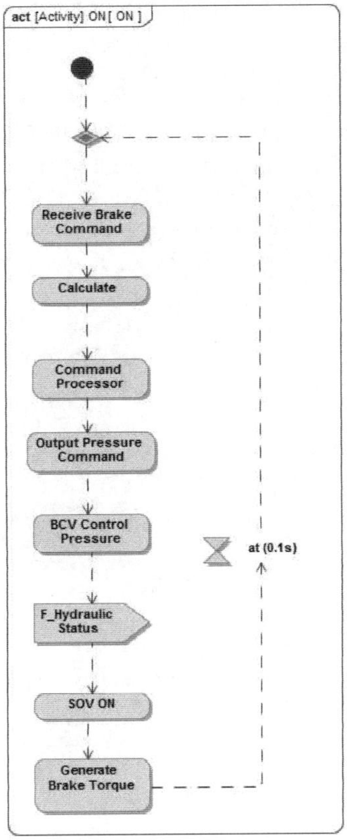

(b) Autobrake ON

Figure 6 ACT

6. Conclusion

MBSE has become the trend in the future aviation field. In this paper, the concept of MBSE is introduced into the functional design of civil aircraft braking system. On the one hand, the functional architecture model is built by combining the system modeling language, and the interaction between functions is completed. On the other hand, the modeling method of MBSE in the logical definition of brake system is determined, and the modeling work is carried out with this method to build the mechanism model. On this basis, this paper defines the inputs of other systems by digital method, realizes the simulation of braking system mechanism, and improves the reliability and feasibility of the model. Of course, the architecture design of aircraft system is a complex problem in many fields. In the future, it is hoped that it can be jointly simulated with other systems to simulate a more realistic braking scene.

References

[1] Zhu J.,Yang H.,Gao Y. h. (2016) Summary of Model Based System Engineering. In: Aeroengine. China. 42(4): 12-16.
[2] INCOSE. Systems Engineering Vision 2025(2014).
[3] Mazeika D., Morkevicius A., Aleksandraviciene A. (2016) MBSE driven approach for defining problem domain. In: 2016 11th System of Systems Engineering Conference.
[4] Arcuti, C., Brusa, E., Maggiore, P., Valfre, E., Mancuso, G. (2019) Design of the Environmental Control System of an Unmanned Aerial Vehicle through the MBSE. In: 2019 International Symposium on Systems Engineering (ISSE). Italy.
[5] Michael, C. (2011). Aircraft Braking Control. EP2035264B1.

[6] Yang, H., Xiao, Y., & Bing, L. I. (2018) Mbse application in civil aircraft brake system requirements analysis. In: Civil Aircraft Design & Research. China. pp. 104-108.
[7] Mei, Q., Huang, D., Yi, L. U. (2019) Design method of civil aircraft functional architecture based on mbse. In: Journal of Beijing University of Aeronautics and Astronautics.,45(5): 1042-1051
[8] Ming, Z., Hong, N., Wei, X. H., Zhou, E. (2009) Research on modelling and simulation for aircraft anti-skid braking. In: Systems and Control in Aerospace and Astronautics, 2008.
[9] Holt, John, S.Perry. (2008) SysML for Systems Engineering. In: Institution of Engineering and Technology.
[10] Xu Z. (2017) Tool Chain Construction of Civil Aircraft Design Based on MBSE Method. In: Aeronautical Manufacturing Technology. China. pp. 100-104.

Shock Train Leading Edge Detection Method under the Condition of Steady Back Pressure and Fluctuating Back Pressure

Mengyang Tian[1, a], Meng Zhao[1, b], Xinyi Liu[1, c], Zhiju Liu[2, d], Yaojia Yang[3, e], Xinyu Liu[4, f], Bing Xiong[5, g*]

[1]Organization: College of Aerospace Science and Engineering, National University of Defense Technology, Changsha, Hunan, China

[2]Organization: College of Aerospace Science and Engineering, National University of Defense Technology, Changsha, Hunan, China

[3]Organization: College of Aerospace Science and Engineering, National University of Defense Technology, Changsha, Hunan, China

[4]Organization: College of Aerospace Science and Engineering, National University of Defense Technology, Changsha, Hunan, China

[5]Organization: College of Aerospace Science and Engineering, National University of Defense Technology, Changsha, Hunan, China

[a]1027476090@qq.com, [b]1290185302@qq.com, [c]2472506294@qq.com, [d]1093898238@qq.com, [e]yangyaojia@nudt.edu.cn, [f]934648182@qq.com

[g*]Corresponding Author: xiongbing10@nudt.edu.cn

Abstract—In this paper, the shock train in the isolator is taken as the research object, and the method of detecting the leading edge position of the shock train during self-excited oscillation and forced oscillation is studied by conducting the wind tunnel test. A steady-state throttling system and a fluctuating throttling system are designed to apply the required steady-state and fluctuating back pressure downstream of the isolator. Then, the four methods of pressure ratio method, pressure difference method, standard deviation method, and Fast Fourier Transform (FFT) analysis method are used to detect and study the position of the shock train leading edge under steady and fluctuating back pressure to verify their applicability. The results show that under the condition of steady-state back pressure, the four methods can get effective results, and the results are consistent. Under the condition of fluctuating back pressure, the pressure difference method could not give effective results, but the other three methods could give effective and consistent results. And the results of the standard deviation method and the Fast Fourier Transform analysis method are similar in image and principle.

1. Introduction

The isolator is an important component of the scramjet. Its main function is to isolate the impact of high back pressure in the combustion chamber on the upstream and to generate a shock train to further pressurize the incoming flow[1]. The structure of the shock train is usually a series of normal shock waves and oblique shock waves, and the flow field is very complex.

To simplify the study, the back pressure of the combustion chamber is often fixed to study the flow field characteristics under steady conditions. At present, great progress has been made, the characteristics of the self-excited oscillation of shock train under steady-state back pressure conditions have been relatively clear, and the main achievements include shock train length prediction, shock train structure characteristics, wall pressure characteristics of the isolator, shock train self-excited oscillation phenomenon, etc. [2]

However, in actual work, there is a combustion oscillation phenomenon[3] in the combustion chamber, which makes the outlet back pressure fluctuate. The upward propagation of this fluctuating back pressure causes the forced oscillation of the shock train. In addition, during the flight of the real aircraft, parameters such as the combustion chamber pressure and the incoming Mach number will inevitably change, which will also cause the shock of the outlet back pressure. The resulting forced oscillations of the shock train can cause problems such as flow separation and non-starting of the intake, etc.[4] Therefore, the detection of the leading edge of the shock train under the condition of fluctuating back pressure and the research on the characteristics of forced oscillation are urgent problems to be solved. The purpose of this paper is to use various methods to detect the leading edge of the shock train under steady-state and fluctuating conditions.

Commonly used methods comprise the pressure ratio method, pressure difference method, standard deviation method, and Fast Fourier Transform (FFT) analysis method[2]. These methods are relatively mature in the study of self-excited oscillation, but their application in the study of forced oscillation is yet to be verified. In this paper, the above four methods are used to detect and analyze the position of the leading edge of the shock train in the two phenomena of self-excited oscillation and forced oscillation.

2. Experiment Apparatus and Data

2.1 Experiment Apparatus

A continuous supersonic wind tunnel of NUDT (National University of Defense Technology) Scramjet Laboratory undertook the relevant experiments in this paper. The wind tunnel is shown in Figure 1. It consists of ①Gas collection section, ②Stabilization section, ③Nozzle, ④Test section, ⑤Transfer section, ⑥Throttle section, ⑦Vacuum tank, ⑧Schlieren equipment, ⑨Motor.

Figure 1. Overall view of the wind tunnel[2]

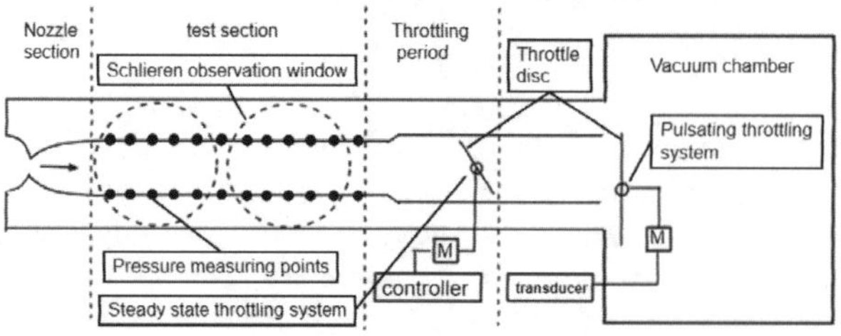

Figure 2. Schematic diagram of the wind tunnel[2]

The test section is a rectangular straight tube of equal section, 400mm long, 56mm high, and 120mm wide. There are 22 pressure measuring holes with a spacing of 15mm on the upper and lower walls. Among them, the coordinates of the high-frequency pressure sensors A_1-A_{12} used in the experiments in this paper are as follows:

Table 1. High-frequency Pressure Sensor A1-A12 Coordinates (mm)

A_1	A_2	A_3	A_4	A_5	A_6	A_7	A_8	A_9	A_{10}	A_{11}	A_{12}
44	74	104	134	164	194	224	254	284	314	344	374

A steady-state throttling system consists of a throttling disc, servo motor, controller, and control computer. The throttling disc is installed in the throttling section, which can realize 360-degree rotation under the drive of the controller and the motor, and the speed and angle are controlled by the computer. The blocking range of the discs is 0-52.14%, thus maintaining a certain steady back pressure.

The pulsation throttling system is installed in the vacuum tank and consists of a throttling rotary vane, a variable frequency motor and a frequency converter. The variable frequency motor drives the throttling vane to rotate, and the area shielded by the vane at the outlet of the wind tunnel changes periodically, which causes fluctuating back pressure. The fluctuating back pressure of 15-300Hz can be realized by adjusting the relative position of the motor and the rotor at the outlet of the wind tunnel, combined with the adjustment of the current frequency by the inverter. It is known that the pressure pulsation frequency in the scram combustion chamber is tens to 200Hz, so the pulsation throttling system fully meets the requirements.

The observation equipment in the experiment is PhotronFastcam SA5 high-speed camera and zigzag schlieren system. The schlieren system can reflect the change of gradient density in the flow field and capture the shock wave in the flow field. The experimental pressure acquisition system consists of an NS-2 high-frequency pressure sensor, data transmission line, and two dynamic test analyzers DH5956 and DH5925.

2.2 Experiment Data

In this paper, data from two sets of experimental conditions is studied. The incoming Ma number of the two experiments is 2.0, but different back pressure conditions are applied. The condition of Experiment 1 is steady back pressure. Under this condition, there exists self-excited oscillation in the shock train. The condition of Experiment 2 is fluctuating back pressure condition, and there exists forced oscillation in the shock train. The experimental conditions are respectively:

2.2.1 Experiment 1. Five steady-state back pressure conditions are applied downstream, and the back pressure sequence is expressed as 29-31-33-34-35, while the number means the angle of the throttling disc, in °. The angle can determine the degree of blockage of the outlet, thus determining the back pressure. The larger the rotation angle is, the greater the back pressure is. Each back pressure is maintained for one second. The sampling frequency is 2000Hz, and the unit of the pressure is Pa. Figure 3 shows the pressure of the sensor A_1-A_{12} in the valid experiment time from 4.4s to 9.5s.

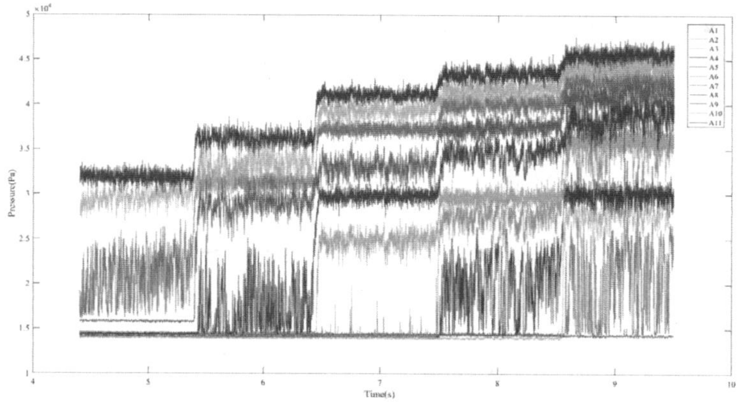

Figure 3. Pressure signal of Experiment 1

2.2.2 Experiment 2. The average value of the back pressure is 4 staircase values, with an incentive frequency of 90Hz, the sampling frequency is 1000Hz, and the unit of the pressure is Pa. Figure 4 shows the pressure of the A_1-A_{12} sensors in the valid experiment time from 0s to 3.7s.

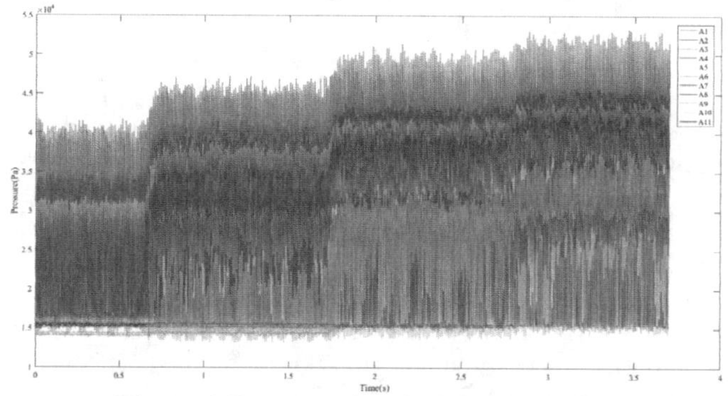

Figure 4. Pressure signal of Experiment 2

3. Leading Edge Detection of Self-excited Oscillation Shock Train Flow Field

3.1 Pressure Ratio Method under Steady Back Pressure

3.1.1 Introduction to the pressure ratio method. The pressure ratio method is a detection method based on the original pressure signal. A detailed introduction to this method can be found in [5]. Its discrimination process is as follows:

$$\mathbf{PR}(k\Delta t) = \left[\frac{p_{s,A1}(k\Delta t)}{p_{tare,A1}}, \frac{p_{s,A2}(k\Delta t)}{p_{tare,A2}}, \cdots, \frac{p_{s,A12}(k\Delta t)}{p_{tare,A12}}\right] \tag{1}$$

$$x_{STLE,PR}(k\Delta t) = interp\left(\mathbf{PR}(k\Delta t), \mathbf{x_{trans}}, PR_{des}\right) \tag{2}$$

In these formulas, $PR(k\Delta t)$ is the vector composed of the pressure ratio at each measuring point, $\mathbf{x_{trans}}$ is the vector consisting of the location of the pressure measuring points, and PR_{des} is the expected pressure ratio, and also the interpolation point to obtain the leading edge position of the shock train. It can be seen from the research in [5] that under the condition of $Ma2$ inflow when the expected pressure ratio PR_{des} is 1.5, the STLE detected by the pressure ratio method is the closest to the test results. Therefore, the results of the pressure ratio method are used in this paper as the reference standard for the results obtained by other methods.

3.1.2 Analysis of the results of the pressure ratio method. Figure 5 shows the detection results of the leading edge of the shock train in the self-excited oscillation mode. The method used is the pressure ratio method, and the desired pressure ratio PR_{des} is taken as 1.5. In the figure, the horizontal axis is the detection time, and the vertical axis is the real-time position of the leading edge of the detected shock train, which is represented by the length (mm) from the entrance of the isolator.

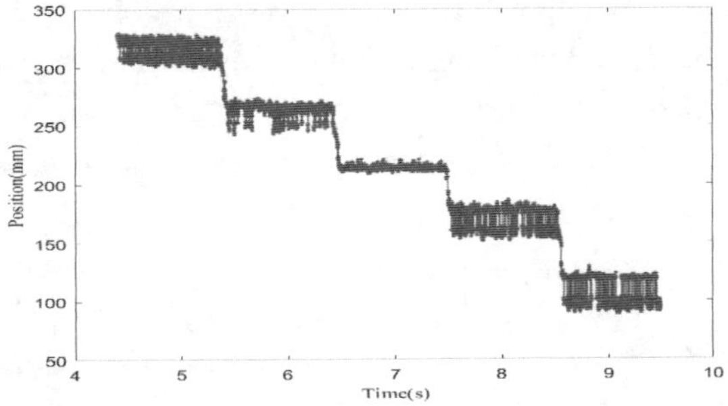

Figure 5. Detection results (self-excited oscillation, pressure ratio method)

A total of 5 steady-state back pressure conditions are applied in this working condition. It can be seen from the figure that when the working condition is Back pressure 1, the time-averaged leading edge position of the shock train is about 320, and then with an increase of back pressure, the time-averaged leading edge position of the shock train gradually moves to the entrance of the isolator (The specific values are shown in Table 2). Finally, when the working condition is Back pressure 5, the time-averaged leading edge position of the shock train reaches about 110.

Table 2. Detection results under each back pressure condition (self-excited oscillation, Pressure Ratio Method)

Working Condition	Time-averaged Leading Edge Position	Time-averaged Amplitude
Back pressure 1	315.67	11.81
Back pressure 2	261.43	12.72
Back pressure 3	214.40	3.32
Back pressure 4	167.58	12.50
Back pressure 5	110.54	14.35

At the same time, it can be found that under each back pressure condition, the position of the leading edge of the shock train oscillates around its time-averaged value. Note that under the condition of Back pressure 3, the time-averaged amplitude detected by the pressure ratio method is very small, but after changing the expected pressure ratio PR_{des} from 1.5 to 1.7, the time-averaged amplitude value of Back pressure 3 changes to 10.37. Therefore, it can be seen that when the desired pressure ratio PR_{des} is set to 1.5, large oscillations may not be detected under certain back pressure conditions. At this time, the desired pressure ratio needs to be adjusted and optimized to obtain a better detection effect.

3.2 Pressure Difference Method under Steady Back Pressure

3.2.1 Introduction to the pressure difference method. The pressure difference method is a method of taking the range based on a sliding time window. A specific introduction to the pressure difference method is given in [2]. Its discrimination process is as follows:

$$d_{Ai,Aj}(k\Delta t) = |p_{Ai}(k\Delta t) - p_{Aj}(k\Delta t)| \tag{3}$$

$$\Delta d_{Ai,Aj}(k\Delta t) = \max(d_{Ai,Aj}((k-l)\Delta t{:}k\Delta t)) - \min(d_{Ai,Aj}((k-l)\Delta t{:}k\Delta t)) \tag{4}$$

In these formulas, $d_{Ai,Aj}(k\Delta t)$ is the absolute value of the real-time pressure difference between measuring point A_i and measuring point A_j, $\Delta d_{Ai,Aj}(k\Delta t)$ is the extreme difference of the pressure difference $d_{Ai,Aj}(k\Delta t)$ within a certain data length, that is, the difference between the maximum pressure difference and the minimum pressure difference. l is the data length, and is taken as 10 in this paper.

The criterion for judging that the leading edge of the shock train reaches the detection point i is shown in (5):

$$\begin{cases} \Delta d_{A1,A12}(k\Delta t) < \alpha \\ \Delta d_{Ai,A12}(k\Delta t) > \beta \quad (i = 2, 3, ..., 11) \end{cases} \tag{5}$$

3.2.2 Analysis of the results of the pressure difference method. Figure 6 shows the detection results of the leading edge of the shock train in the self-excited oscillation mode. The method used is the pressure difference method, and the meaning of the horizontal and vertical coordinates in the figure is the same as that in Figure 5.

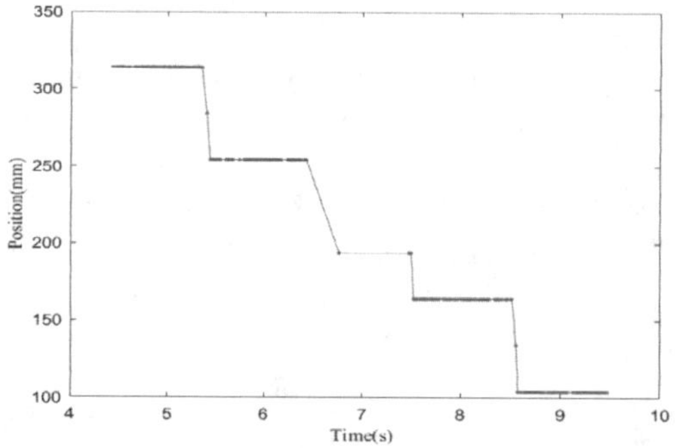

Figure 6. Detection results (self-excited oscillation, pressure difference method)

A total of 5 steady-state back pressure conditions are applied in this working condition. It can be seen from the figure that the position of the leading edge of the detected shock train is stepped, and the values of each segment are shown in Table 3. It can be seen that with an increase of back pressure, the average position of the leading edge of the shock train gradually moves towards the entrance of the isolator, which is consistent with the conclusion obtained by the pressure ratio method. However, due to the characteristics of the method, the pressure difference method does not detect the self-excited oscillation traits of the leading edge of the shock train.

Table 3. Detection results under each back pressure condition (self-excited oscillation, pressure difference method)

Working condition	Time-averaged leading edge position
Back pressure 1	314
Back pressure 2	254
Back pressure 3	194
Back pressure 4	164
Back pressure 5	104

3.2.3 Comparison of the results of the pressure difference method and the pressure ratio method.
Under the same back pressure condition, the time-averaged leading edge position of the shock train detected by the pressure ratio method is set as x_r, with the position detected by the pressure difference method set as x_d, and the distance between two adjacent sensors set as D, the error defined as $\Delta = |x_r - x_d|$. Then when $\Delta \leq D$ is considered that the detection results of the pressure ratio method and the pressure difference method are consistent.

The data is substituted in $D = 30mm$ as well as Table 2 and Table 3 into the check calculation, we can get that under each back pressure condition, the detection results of the pressure ratio method and the pressure difference method are consistent.

3.3 Standard Deviation Method under Steady Back Pressure
The standard deviation method is a detection method based on the dimensionless standard deviation between the sliding time window. The definition of the dimensionless standard deviation is $\sigma\big((k-N+1)\Delta t\colon k\Delta t\big) = \sqrt{\frac{\sum_{i=1}^{N}[(p_i - p_{av})/p_{ref}]^2}{N-1}}$. Among them N is the length of the sliding time window, Δt is the time interval of two data, $k\Delta t$ is the time of interest, *and* p_i is the pressure of one of the pressure sampling points in the time window. p_{av} is the average pressure during this time, and p_{ref} is the reference pressure, which is the upstream pressure, 13500Pa. After experimental verification, three detection standards are proposed. One of the commonly used standards is the maximum standard deviation[6]. This paper selected this standard as a judgment.

The data of Experiment 1 is processed and analyzed. $N = 25$ is taken, and the time and space distribution of $\sigma((k-N+1)\Delta t\colon k\Delta t)$ is calculated to detect the leading edge of the shock train according to the maximum standard deviation. Meanwhile, compared with the results of the pressure ratio method, the results are shown in Figure 7.

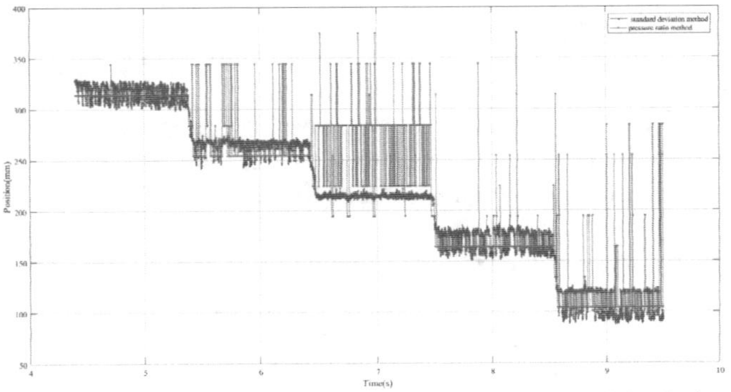

Figure 7. Detection results (self-excited oscillation, standard deviation method)

In Figure 7, "Time" represents the time of the experiment. "Position" represents the distance from the leading edge of the shock train to the entrance of the isolator.

The time-averaged leading edge position of the shock train is calculated under the five types of back pressure, as shown in Table 4.

Table 4. Detection results under each back pressure condition (self-excited oscillation, standard deviation method)

Working Condition	Time-averaged Leading Edge Position
Back pressure 1	314.06
Back pressure 2	264.97
Back pressure 3	262.60

Back pressure 4	165.44
Back pressure 5	113.98

In addition, to further observe the distribution characteristics of the standard deviation under different back pressure conditions and compare the results with the pressure ratio method, the typical standard deviation spatial distribution under five different back pressure conditions and the corresponding detection positions of the pressure ratio method at the time $k\Delta t$ are drawn, as shown in Figure 8.

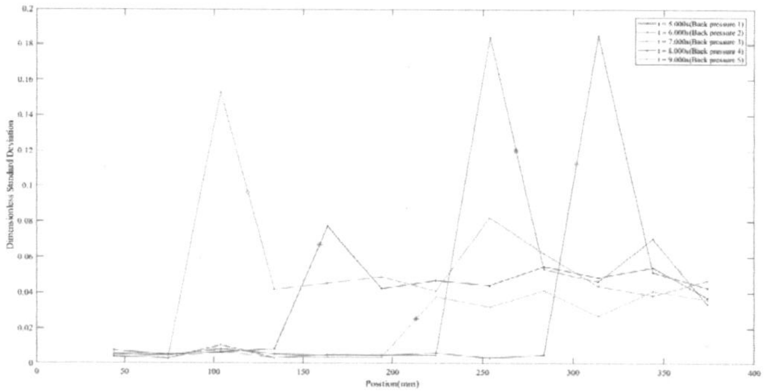

Figure 8. Five typical standard deviations at different back pressure (self-excited oscillation)

In Figure 8, the horizontal axis represents the length from the entrance of the isolator, the vertical axis represents the dimensionless standard deviation, the five curves represent the typical dimensionless standard deviation distribution under five back pressure conditions respectively, and the abscissa of each asterisk on each curve represents the detection value of the pressure ratio method at the time corresponding to the curve.

By the above data process, the following analysis and conclusions can be made for steady back pressure conditions with Ma =2:

1) From Figure 8 it can be inferred that the pressure oscillation upstream of the leading edge is small, so the standard deviation is small. Standard deviations increase near the leading edge of the shock train. In all cases except the third back pressure condition, the standard deviation has a significant decrease in the standard deviation after the sudden increase.

2) From Figure 7 and Table 4, it can be inferred that except for the third back pressure condition, the detection results of the standard deviation method are basically enveloped by that of the pressure ratio method, and the average value of them is close, which proves that the results of the two detection methods are consistent to a certain extent.

3) From Figure 7, Figure 8, and Table 4, it can be concluded that under the third back pressure, the shape of the standard deviation distribution is significantly different from other back pressure conditions, and the standard deviations near the leading edge and downstream of the shock train are close, resulting in large fluctuations in detection results. One possible reason is that the leading edge of the shock train is located in the middle of the two measuring points, which causes the result of the relatively low maximum value of the standard difference. However, by observing the high standard deviation distribution length under other back pressure conditions, it can be inferred that there might be other reasons.

3.4 FFT Analysis Method under Steady Back Pressure

Under the five steady-state back pressure conditions of Experiment 1, the pressure data of 300 time points under the back pressure condition is selected from 12 pressure measuring points to perform a fast Fourier transform. The processing and analysis results of the first two back pressure conditions are as follows:

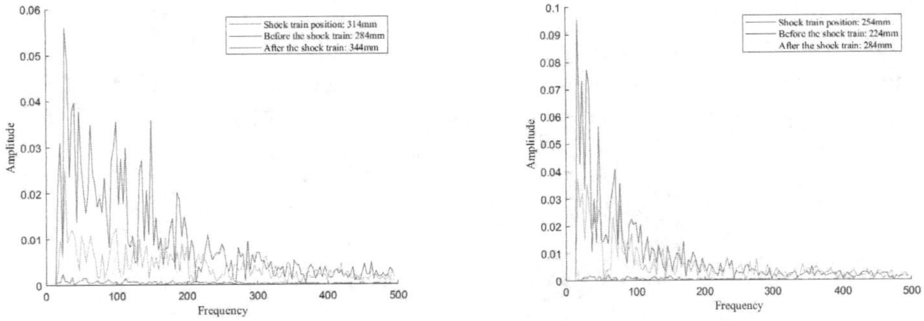

Figure 9. FFT of self-excited oscillation (Back pressure 1 and 2)

It can be seen from the figure that under the first back pressure condition, there is a curve with the highest amplitude (blue curve), which corresponds to the measuring point with a position of 314mm. Compared with the pressure ratio method, it can be found that under this back pressure condition The position of the leading edge of the shock train is the measurement point at 314mm, from which it can be inferred that the measurement point corresponding to the curve with the highest FFT amplitude is the position of the leading edge of the shock train.

Under the second back pressure conditions, compared with the pressure ratio method, the FFT graph also proves this point. Only the FFT graphs under the first two back pressure conditions are given here.

Further, the positions of the upstream and downstream measuring points of the shock train leading can be determined, which are represented by the red and yellow curves, respectively.

Finally, the Fourier transform method is summarized: under the condition of steady back pressure, the pressure values of all pressure measuring points within a period of time are Fourier transformed, and in the obtained FFT amplitude map, the pressure measuring points corresponding to the highest amplitude curve is the position of the leading edge of the shock train. Here, the max average FFT amplitude can be used to indicate the highest amplitude curve.

The average FFT amplitude of each pressure measurement point under five back pressure conditions is plotted as follows. It can be seen that the average value of the amplitude upstream of the shock train leading position is small and close to zero, while the amplitude increases sharply at the leading edge shock train position. The obtained results are the same as those of the pressure ratio method, which proves that the FFT method is feasible under self-excited oscillation.

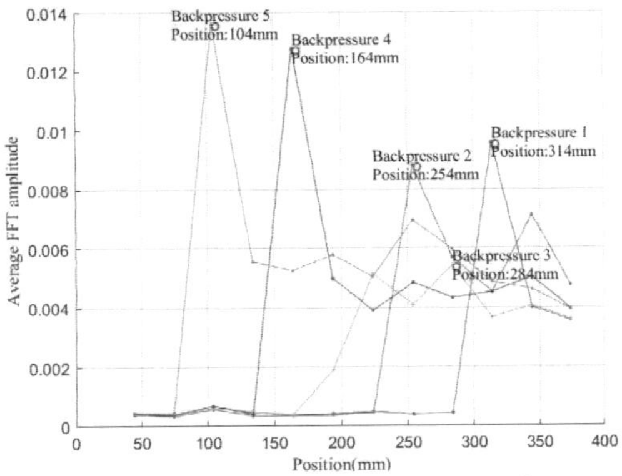

Figure 10. Average FFT amplitude under five back pressure conditions

Comparing Figure 10 with the standard deviation method (Figure 8), it can be found that the trend is similar, which reflects the similarity in principle between the two methods, that is, the consistency of standard deviation and FFT amplitude in characterizing the position of the leading edge of the shock train.

4. Leading Edge Detection of Forced Oscillatory Shock Train Flow Field

4.1 Pressure Ratio Method under Fluctuating Back Pressure

Using the same method as in the self-excited oscillation case, the desired pressure ratio PR_{des} is taken as 1.5, the following results can be obtained after processing the data of Experiment 2.

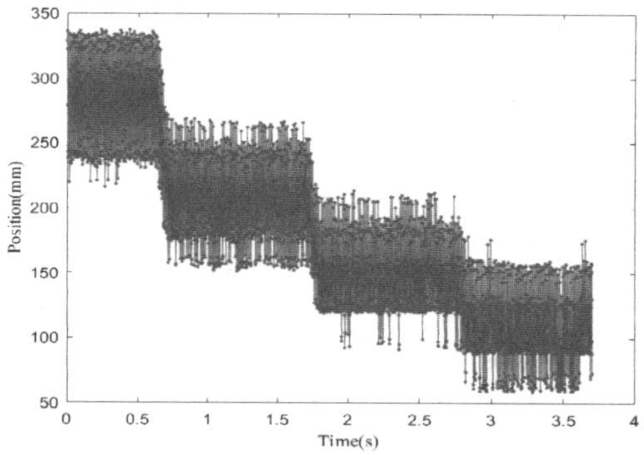

Figure 11. Detection results (forced oscillation, pressure ratio method)

Figure 11 shows the detection results of the leading edge of the shock train in the forced oscillation mode. The method used is the pressure ratio method, and the meaning of the horizontal and vertical coordinates in the figure is the same as that in Figure 5.

In this working condition, a back pressure with an incentive frequency of 90Hz is applied, and the average value of the back pressure is four staircase values. The time-averaged leading edge position of the shock train under each back pressure condition is shown in Table 5. It can be seen that the time-averaged leading edge position of the shock train is negatively correlated with the back-pressure time-averaged value, that is, when the back-pressure time-averaged value increases, the shock train leading edge tends to move toward the entrance, and the position coordinate average value decreases.

Table 5. Detection results under each back pressure condition (forced oscillation, pressure ratio method)

Working Condition	Time-averaged Leading Edge Position	Time-averaged Amplitude
Back ressure 1	286.52	42.20
Back pressure 2	216.70	29.68
Back pressure 3	162.46	31.42
Back pressure 4	127.65	30.22

In addition, by comparing Table 2 and Table 5, it can be obtained that the time-averaged amplitude under forced oscillation is significantly larger than that under self-excited oscillation.

4.2 Pressure Difference Method under Fluctuating Back Pressure

Using the same method as in the self-excited oscillation case, taking $\alpha=1.5$, $\beta=0.5$, the following results can be obtained after processing the data of Experiment 2.

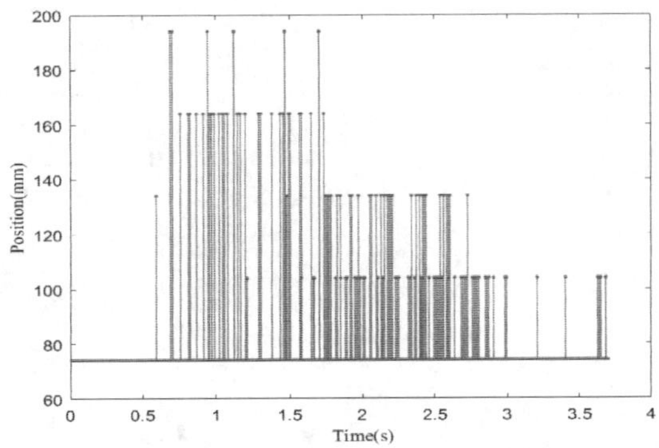

Figure 12. Detection results (forced oscillation, pressure difference method)

Figure 12 shows the detection results of the leading edge of the shock train in the forced oscillation mode. The method used is the pressure difference method, and the meaning of the horizontal and vertical coordinates in the figure is the same as that in Figure 5.

It can be seen from Figure 11 and Figure 12 that the results are quite different from the detection results of the pressure ratio method, which are obtained by using the pressure difference method introduced in this paper to detect the position of the leading edge of the shock train under the condition of forced oscillation. The results shown in Figure 12 have no regularity, and the detection effect is very poor. The pressure difference method under the condition of forced oscillation still has room for exploration and improvement.

4.3 Standard Deviation Method under Fluctuating Back Pressure

Use the same method as in the self-excited oscillation case to process the data of Experiment 2. $N = 25$ is taken, and the time and space distribution of $\sigma((k - N + 1)\Delta t : k\Delta t)$ are calculated to detect the leading edge of the shock train according to the maximum standard deviation. After comparing with the results of the pressure ratio method, the results are shown in Figure 13.

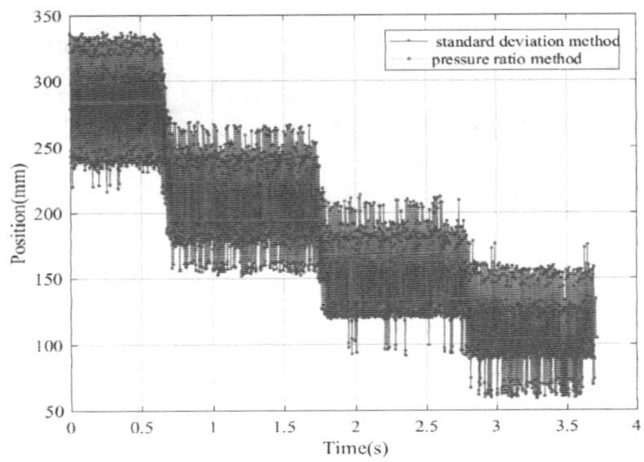

Figure 13. Detection results (forced oscillation, standard deviation method)

Then, the time-averaged leading edge position of the shock train is calculated under five types of back pressure, as shown in Table 6.

Table 6. Detection results under each back pressure condition (forced oscillation, standard deviation method)

Working Condition	Time-averaged Leading Edge Position
Back pressure 1	284.00
Back pressure 2	203.13

Back pressure 3	154.42
Back pressure 4	107.85

Four typical standard deviation space distributions are drawn under different back pressure conditions and the corresponding pressure ratio detection position is drawn at time $k\Delta t$, as shown in Figure 14.

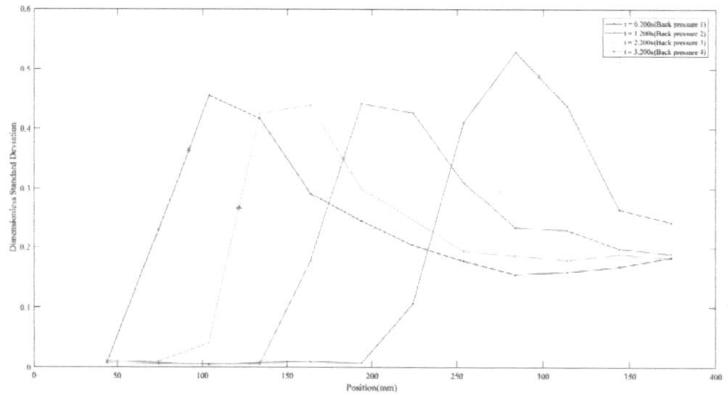

Figure 14. Four typical standard deviations at different back pressure (forced oscillation)

By the above data process, the following analyses and conclusions can be made for fluctuating back pressure conditions with Ma =2:

1) It can be seen from Figure 13 that the detection results of the standard deviation method are basically enveloped by that of the pressure ratio method, and the time-averaged values of the two methods are close, which proves that the results of the two detection methods under the fluctuating back pressure are consistent to a certain extent.

2) It can be seen from Figure 14 that the standard deviation distribution under fluctuating back pressure conditions is similar to the shape in steady back pressure conditions, but the magnitude of the mutation of the standard deviation and the distance spanned by the mutation are both higher than those of the steady back pressure conditions. It may be resulting from the amplitude and frequency of the forced oscillation under the fluctuating back pressure are larger than that of the self-excited oscillation.

3) From the comparison of Figure 13 and Figure 7, it can be seen that under the fluctuating back pressure, the detection result of the standard deviation method oscillates less, which implies that this method may not be suitable for detecting the forced oscillation characteristics of the shock train.

4.4 FFT Analysis Method under Fluctuating Back Pressure

Fast Fourier analysis is performed on the five back pressure conditions of the forced oscillation experiment under fluctuating back pressure (time point = 300), and the results for the first two back pressure conditions are taken as shown in the figure below.

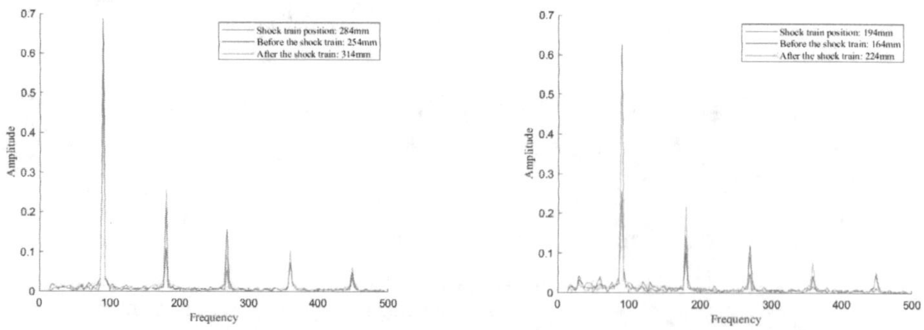

Figure 15. FFT of forced oscillation (Back pressure 1 and 2)

It can be seen that, compared to the FFT amplitude image of the self-excited oscillation, the image of the forced oscillation exhibits five successively decreasing peaks. Compared with the results of the pressure ratio method, it is found that the measurement point corresponding to the curve with the largest peak (blue curve) is the position of the leading edge of the shock train.

The average FFT amplitude of each pressure measurement point under the four back pressure conditions is plotted as follows. It can be seen that the surge in amplitude is the position of the leading edge of the shock train, which is the same as the result measured by the pressure ratio method, which proves that the FFT method is feasible under forced oscillation.

Similarly, under the condition of fluctuating back pressure, the trend of the image below is similar to that of the standard deviation method (Figure 16).

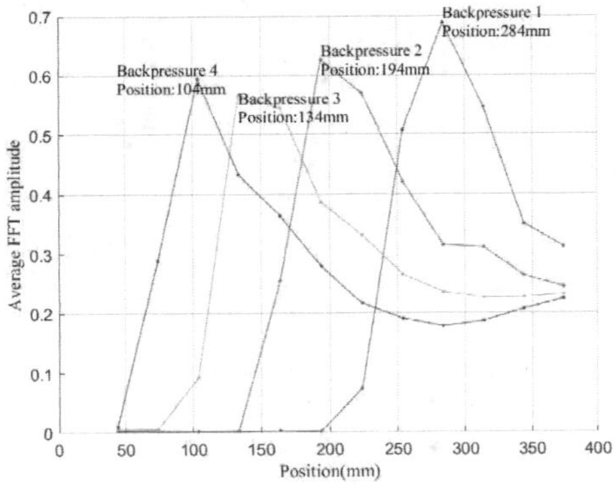

Figure 16. Average FFT amplitude under four back pressure conditions

5. Conclusion

In this paper, the pressure ratio method, the pressure difference method, the standard deviation method, and the Fast Fourier Transform (FFT) analysis method are used to detect the leading edge of the shock train under the steady back pressure and the fluctuating back pressure, and the detection results of each method are compared and analyzed. The fit and validity of the results are as follows:

1) The four methods apply to the self-excited oscillation of the shock train, and the obtained results are consistent, which verifies the applicability of the four methods under the condition of steady back pressure.

2) The pressure ratio method, the standard deviation method, and the Fast Fourier Transform (FFT) analysis method are applicable under fluctuating back pressure, and the trends of the average value are consistent, but the pressure difference method cannot obtain effective results.

3) Among the methods mentioned, the pressure ratio method could be considered as the most possible method to detect the self-excited oscillation and forced oscillation of the shock train, while the other methods can only detect the time-averaged result to a certain extent, which is decided by how much historical data is used.

Acknowledgments

The authors gratefully acknowledge the support of the National Natural Science Foundation of China (Grant No. 12102470) and the NUDT Research Project (No. ZK20-05) for this work.

References

[1] Xiao Yinli,Chen Liang,Song Wenyan.Study on flow characteristics of scramjet isolated section[J].Acta Aerodynamics,2007(01):75-79.
[2] Xiong Bing. Study on the forced oscillation characteristics of shock trains in the isolator[D]. National University of Defense Technology, 2016.
[3] Ma F H, Li J, Yang V. Thermo-acoustic flow instability in a scramjet combustor [C]. AIAA Paper, 2005-3824,2005.

[4] WagnerJ L, Yuceil K B, Valdivia A, Clemens N T, Dolling D S.Experimental Investigation of Unstart in an Inlet/Isolator Model in Mach 5 Flow[J].AIAA Jounal,2009,47(6):1528-1542.
[5] Hutzel J R, Decker D D, Cobb R G, King P I, Veth M J. Scramjet isolator shock train location techniques [C]. AIAA Paper 2011-402, 2011.
[6] Le D B, Goyne C P, Krauss R H. Shock Train Leading-Edge Detection in a Dual-Mode Scramjet [J].Journal of Propulsion and Power,2008,24 (5):1035-1041.
[7] Zhan Wangjie, Yan Hong. Numerical study on the effects of steady-state and fluctuating back pressure on shock trains in isolator [J]. Propulsion Technology, 2021.
[8] Wang Hongbo. Stable jet combustion mode and oscillation mechanism of the concave cavity in supersonic airflow [D]. Changsha: National University of Defense Technology, Ph.D., 2012.

Design and Research of Test Bench for Loading and Performance Testing of Aerospace Precision Ball Screw

Yang Zhang[a]*, YuanXun Fan[b], LuHui Xu

School of mechanical engineering, Nanjing University of Science and Technology, Nanjing, Jiangsu, China

[a*]email: zhangyang08alan@163.com; [b]email: fanyx@mail.njust.edu.cn

Abstract. Aiming at the problem of insufficient research on the motion failure caused by the degradation of the grease of the precision ball screw pair under the typical working conditions of high frequency and high load in aerospace, a set of high frequency inertial loading test bench is designed. Loading and performance testing functions are integrated. The mechanical structure of the test bench, the inertia adjustment device and the connecting member of the pendulum rod are designed, which can simulate the loading under aerospace conditions. Installing multiple sensors can test the force response characteristics, vibration characteristics, temperature rise characteristics and transmission efficiency of the precision ball screw pair when it operates under different frequencies and different axial forces. The axial force of the nut with different rotational speeds under inertial load is tested by experiments.

1. Introduction

Precision ball screw pair has the characteristics of high precision, high load, high efficiency, et al., and plays an important role in the aerospace field, such as rocket nozzle vector control system and aircraft attitude control system [1-4]. Zhang Xiangdong et al. built a vibration signal acquisition test system, which can simulate the poor lubrication of the ball screw under machine tool conditions [5]. Cao Lei et al. designed a friction torque measurement test bench, which can achieve accurate measurement of torque by designing a PFTN unit and torque meter [6]. Zhao Lianchun et al designed a vibration measurement and analysis system to study the effect of lubrication on the vibration characteristics of ball bearings [7]. Kwang-Je Oh designed a friction torque test bench to verify the friction torque model [8]. Based on the Stribeck friction effect, Zhou Yuqing et al. used the variable amplitude and variable speed triangle wave test curve to identify the lubrication state of the machine tool feed axis [9]. Antonio Carlo Bertolion studied the no-load drag torque of ball screws under different lubrication conditions through MBD simulation [10]. However, the ball screw pair used in aerospace is a short-term and high-efficiency mechanism, and it is left in the state most of the time. There is a lack of research on the influence of the degradation of the ball screw pair on the transmission performance under aerospace conditions. This paper designs a set of high-frequency inertial loading test bench integrating loading and performance testing functions with multi-sensor technology to provide a test basis for theoretical derivation.

2. Design principle of Test bench

2.1. The performance indicators of test bench

The main indicators and design requirements of the test bench are as follows:

(1) The maximum test speed is 3000r/min, the maximum axial load is 14KN, and the maximum

loading frequency is 12Hz.

(2) Maximum inertial loading is possible:6.7Kgcm^2.

(3) Vibration acceleration sensor detection accuracy:1mv/ms^{-2}.

(4) The temperature rise can be automatically monitored within the range of 200°C for 1 hour and the detection accuracy ≤ 0.1℃.

(5) Mechanism working noise detection: adopt physical shielding structure and frequency filtering to collect the internal movement sound of the mechanism, detection accuracy: 50mv/Pa.

2.2. Structure of test bench

Figure 1 is the front view of the high-frequency inertial loading test bench. The test bench consists of a drive motor, ball screw pair, acceleration sensor, torque and speed sensor, data acquisition card, temperature sensor, noise sensor, linear guide rail, tension pressure sensor, magnetic powder brake, et al. The console consists of an upper computer, real-time controller and magnetic powder brake controller, et al.; the drive is mainly composed of a motor and driver.

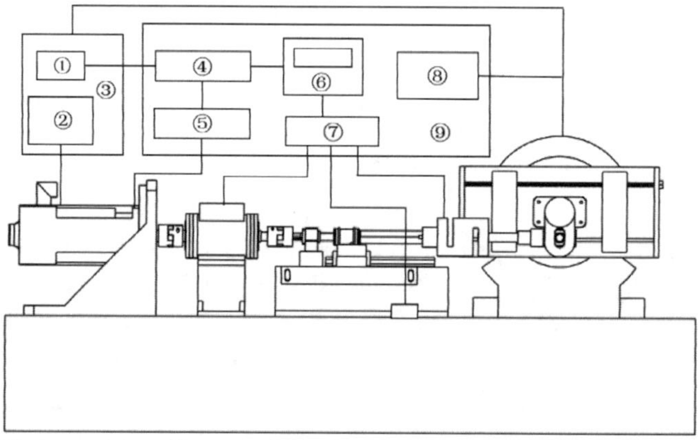

①Power supply;②Motor drive;③Drive;④Real-time controller;⑤Motion control card;⑥Upper computer;⑦Data acquisition card;⑧Magnetic Powder Brake Controller;⑨Console

Figure 1. Test bench schematic

As shown in Figure 2, the motor at the leftmost end of the test bench provides power, the right end of the servo motor is connected to the left end of the torque and speed sensor through a coupling, and the right end of the torque and speed sensor is connected to the screw through a coupling, and the nut end of the ball screw is connected to the tension/compression force sensor, and the right end of the tension/compression force sensor is connected with the loading device, including the pendulum connection mechanism, adjustable inertia device and magnetic powder brake, et al.

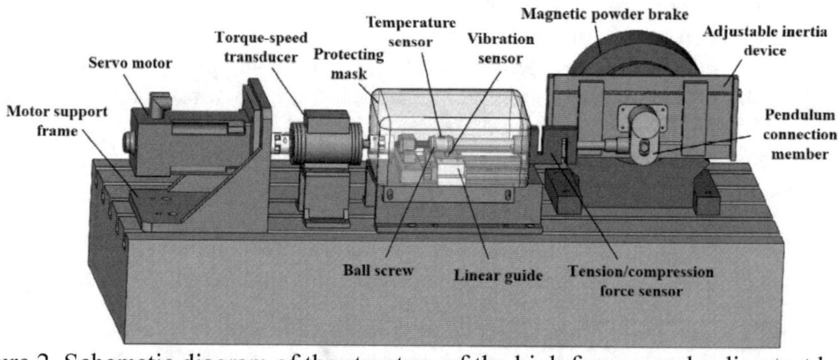

Figure 2. Schematic diagram of the structure of the high frequency loading test bench

2.3. Component selection

Table 1. Component selection

Components	Model	parameter
Motor	AKM64Q	5000rpm 10.2Nm 5.6kW
Magnetic powder brake	CZ-30	300Nm
Torque-speed transducer	ZH07-20	20Nm 6000rpm
Tension/compression force sensor	PPM232-XT-4	5000KN
Vibration sensor	DH311E	5000m/s^2
Temperature sensor	PT100	-50~200℃

3. Design of key parts

3.1. Design of adjustable inertia device

In order to realize the function of inertia loading, an adjustable inertia device is designed, which can quickly adjust the inertia. As shown in Figure 3, the device consists of a screw rod, two mass blocks, a guide bar and a fixed outer frame. The left and right sides of the screw rod are threads with opposite directions and the same pitch. By adjusting the nut to rotate the screw rod, the position of the two mass blocks from the centre can be adjusted at the same time, and the rotation radius can be changed, to realize the adjustment of inertia.

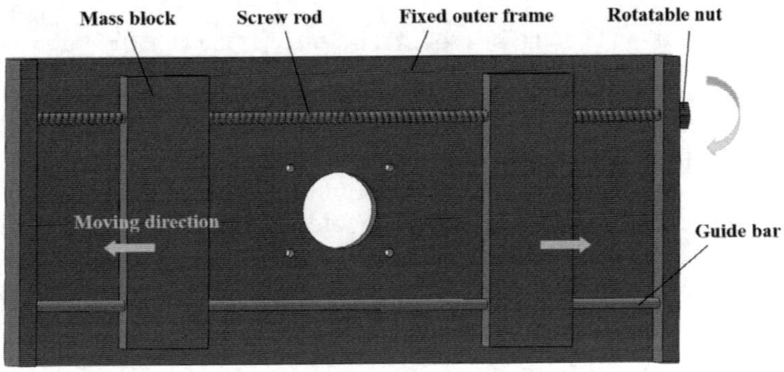

Figure 3. Adjustable inertia device

3.2. Design of swing rod connecting device

As shown in Figure 4, in order to convert the torque provided by the magnetic powder brake into the axial force required when the ball screw is loaded, a swing rod connection device is designed for conversion. The pusher relates to the pendulum through a pin, and the pendulum relates to the rotating shaft of the magnetic powder brake through a key. The adjustable inertia device is mounted on the pendulum through a nut.

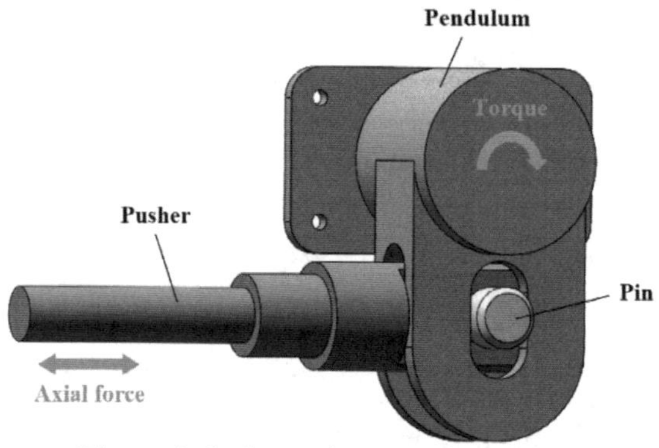

Figure 4. Swing rod connection device

4. Modal analysis

Set the dynamical differential equation for a Linear Time-invariant System:

$$[M]\{\ddot{x}\}+[C]\{\dot{x}\}+[K]\{x\}=\{F(t)\} \tag{1}$$

Among them, [M] is the mass matrix; [C] is the damping matrix; [K] is the system stiffness matrix; x(t) is the displacement vector; x(t) is the velocity vector; x(t) is the acceleration vector; F(t) is the excitation load vector.

Performing Laplace transform on Formula (1), we get:

$$\{s^2[M]+s[C]+[K]\}X(s)=F(s) \tag{2}$$

Considering only the real mode, s can be replaced by $j\omega$ to obtain:

$$\{[K]-\omega^2[M]+j\omega[C]\}X(\omega)=F(\omega) \tag{3}$$

Set $X(\omega)= [\Phi][Q]$, where $[\Phi]= [\Phi_1, \Phi_2, \Phi_3 \ldots \Phi_n]$ is modal matrix, $\Phi_n= [\phi_1, \phi_2, \phi_3 \ldots \phi_i]$, ϕ_i is mode factor, $Q = [q_1(\omega), q_2(\omega), q_3(\omega) \ldots q_n(\omega)]^T$ is modal coordinate vector.

The effects of damping and applied loads can be ignored in the modal analysis of the system, set $[C] = 0$, $F(t) = 0$:

$$\{[K]-\omega^2[M]\}[\Phi]=0 \tag{4}$$

The characteristic equation of Formula (4) is $\omega=\sqrt{K/M}$, which is the natural frequency of the system. Corresponding eigenvectors $[\Phi]$ is the system vibration mode.

Common modal decomposition methods include Subspace iteration method and Lanczos algorithm. In this paper, the Subspace iteration method is used to solve the problem, as shown in Figure 5.

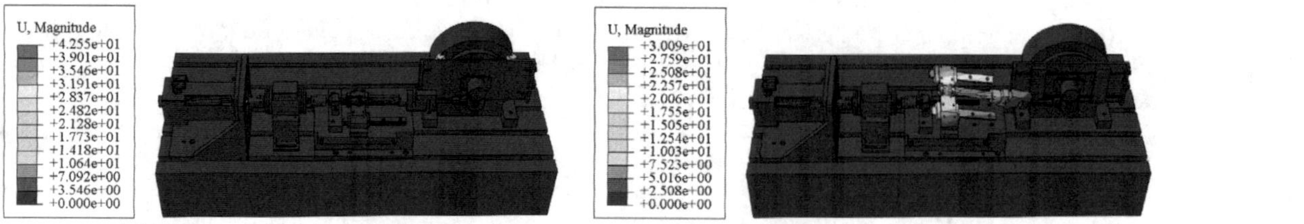

(a) First order mode(68.3Hz) (b) Second order mode(71.873Hz)

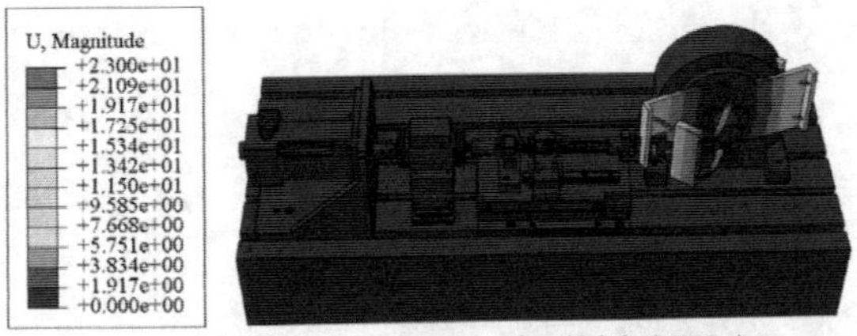

(c) Third order mode (128.07Hz)

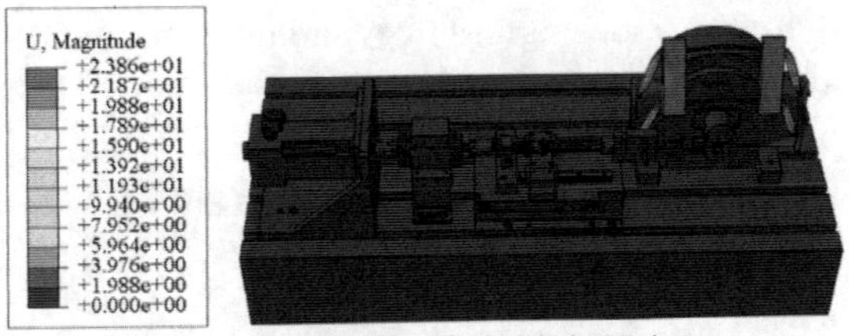

(d) Fourth order mode(193.25Hz)

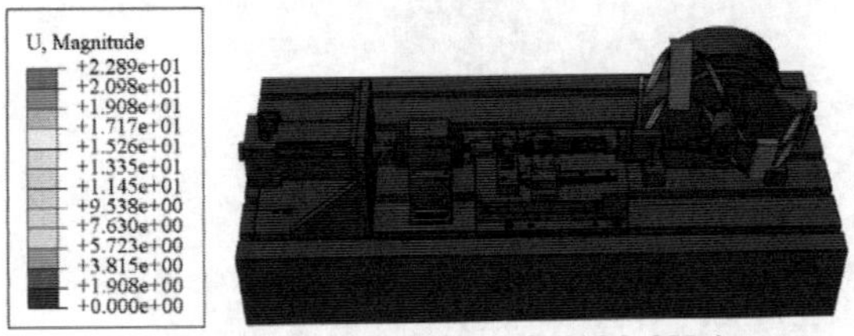

(e) Fifth order mode(199.8Hz)

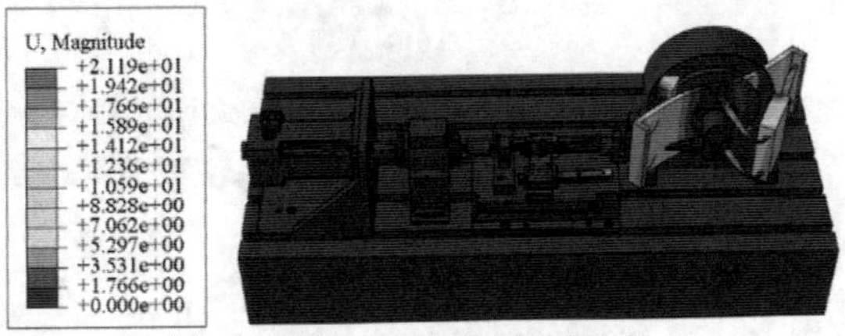

(f) Sixth order mode(202.07Hz)

Figure 5. Test bench modes and their characteristic frequencies

According to Figure 5, the vibration-sensitive area of the test bench is mainly near the inertial loading device, ball screw and motor. The first-order mode is dominated by the vibration of the attachment of the inertial adjustment device, and the natural frequency is 68.3Hz, which is quite different from the 12Hz in the test and will not cause resonance. In the second-order mode, the vibration mainly occurs in the ball screw attachment, which may have a certain impact on the detection of vibration characteristics. It is necessary to pay attention to the test.

5. Experiment of High Frequency Inertia Loading

5.1. Calculation of theoretical inertial force

Let the motor speed:

$$n = n_{\max} \sin[(2\pi f)t] \tag{5}$$

The linear movement speed of the nut is

$$v = P \cdot n_{\max} \sin[(2\pi f)t] \tag{6}$$

Then the linear acceleration of the nut is

$$a = 2\pi fP \cdot n_{\max} \cos[(2\pi f)t] \tag{7}$$

The angular acceleration of the magnetic powder brake is

$$\alpha = \frac{a}{l} = \frac{2\pi fP \cdot n_{\max} \cos[(2\pi f)t]}{l} \tag{8}$$

Then the axial inertia force of the ball screw is

$$F = \frac{J\alpha}{l} = \frac{2\pi fJP \cdot n_{\max} \cos[(2\pi f)t]}{l^2} \tag{9}$$

Where n is the motor speed (r/s), P is the lead of screw(m), f is the frequency of loading (Hz), t is the time(s), l is the distance of force arm, J is rotational inertia (Kgm^2).

5.2. High frequency inertial force loading test

In order to verify the effectiveness of the above high-frequency loading test bench for ball screws, a high-frequency inertia force loading experiment was conducted on a ball screw with matrix return for an aerospace model, as shown in Figure 6.

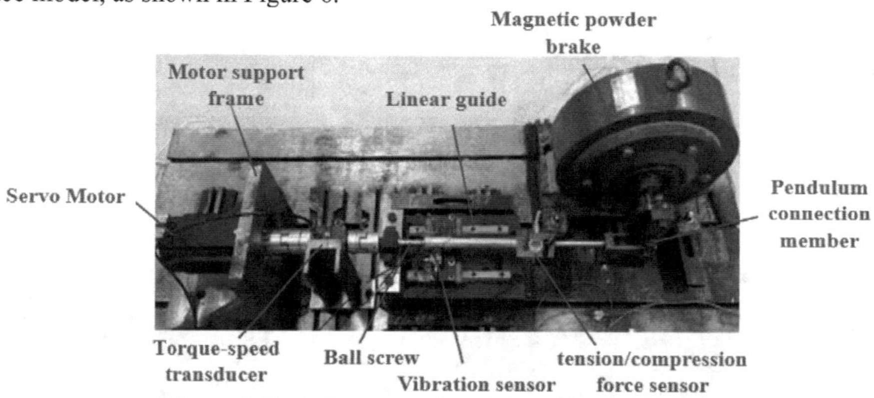

Figure 6. High frequency inertial loading test bench

In the high-frequency inertial force loading experiment, the speed of the motor is set as a sine wave, the frequency is 10 Hz, and the peak speed of the lead screw is 1200 r/min, 1500 r/min and 1800 r/min, and the moment of inertia is 1.1 kgcm^2.

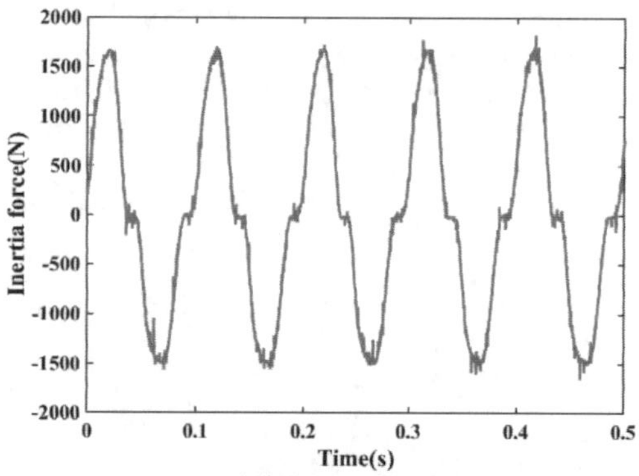

(a)10Hz-1200r/min

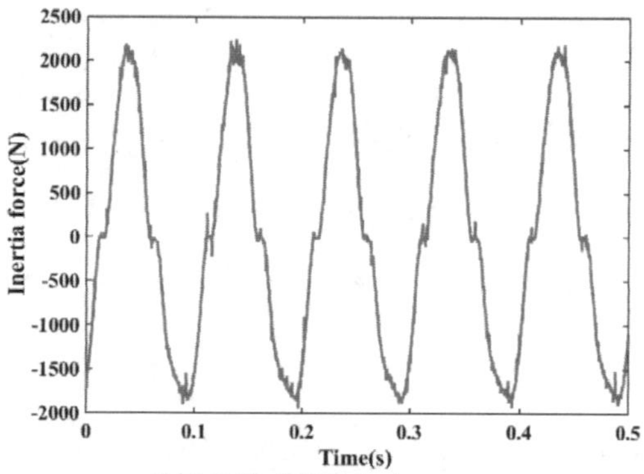

(b)10Hz-1500r/min

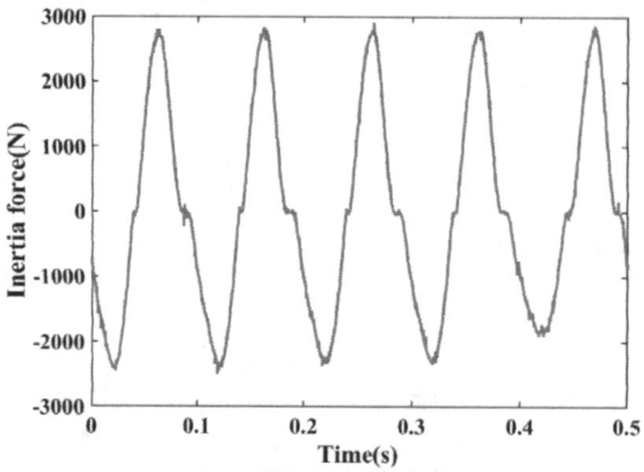

(c) 10Hz-1800r/min

Figure 7. Axial force curve at high frequency

From the test data, the axial force applied to the ball screw during operation increases as the speed becomes larger. During the operation, there are many non-linear factors influencing the operation, such as the clearance between the loading components and the internal resistance torque of the ball screw.

6. Conclusion
The structure of the aerospace high-frequency and high-load test bench is designed, which provides the experimental basis for the performance test of the aerospace servo system. An inertia adjusting device is designed, which can quickly adjust the inertia by turning the nut to achieve inertia loading. A pendulum linkage mechanism is designed to convert the torque provided by the magnetic powder brake into the axial force required when the ball screw is loaded. The modal analysis of the test stand was carried out, and the first 6 orders of vibration frequencies as well as vibration patterns were listed. The axial force of the nut with different rotational speeds under inertial loading is tested through experiments, and the fluctuations caused by the internal resistance torque of the ball screw can be seen.

References
[1] Shi Chao. (2015) Design of Servo Loading Test System for Ball Screw Subs. Nanjing University of Science and Technology.
[2] Sun Baoyuan, Ren Zongjin, Qian Ming, Zhang Jun. (2008) Modeling Analysis of Orbit/Attitude Control Rocket Engine Thrust Dynamic Test System. Journal of Dalian University of Technology,2008(05):668-672.
[3] Peng Qianlong. (2017) Design and Realization of Comprehensive Performance Measurement and Control Platform of Servo Mechanism. Nanjing University of Science and Technology.
[4] Huang Yuping, Li Jianming, Zhu Chenglin. (2020) Aerospace electromechanical servo system. China Electric Power Press, Beijing.
[5] Zhang Xiangdong, Zhou Changguang, Feng Hutian, Ou Yi. (2021) Fault Diagnosis of the Lubrication of Ball Screws Based on KPCA and SVM. Modular Machine Tool & Automatic Manufacturing Technique.2021(05):47-52.
[6] Lei Cao, Kwang-Je Oh, Sung-Chong Chung. Explicit precision friction torque model of ball screws in high-speed operations. Tribology International.152.
[7] Zhao Lianchun, Ma Jiaju, Ma Chunqing, Qiu Zhiren. (2003) Influence of Lubrication on Vibration Characteristics of Ball Bearings. Tribology.23(5),421-425.
[8] Kwang-Je Oh, Lei Cao, Sung-Chong Chung. (2020) Explicit modeling and investigation of friction torques in double-nut ball screws for the precision design of ball screw feed drives. Tribology International.141.
[9] Zhou Yuqing, Mei XueSong, Jiang Gedong, Sun Nuogang, Tao Tao. (2009)Rapid Identification for Feed—axis Lubrication Condition of Large Scale Numerical Control Machine Tools. China Mechanical Engineering.20(2),154-158.
[10] Bertolino Antonio Carlo, Jacazio Giovanni, Mauro Stefano, Sorli Massimo. (2021) Investigation on the ball screws no-load drag torque in presence of lubrication through MBD simulations. Mechanism and Machine Theory.161,157-177.

Research on Static Test Verification Technology of the Connection Between APU and Fuselage of Civil Aircraft

Zhengyi Zhu[1*]

[1]Shanghai Aircraft Design And Research Institute, Shanghai 201210, China

[*]Corresponding author's e-mail: zhuzhengyi@comac.cc

Abstract: This paper studies the static strength verification of the connection between APU and fuselage of large civil aircraft, which provides a verification method for the airworthiness verification of the connection between APU and fuselage. Based on the internal force analysis results of APU finite element, this paper selects the static test load, completes the design from the theoretical load to the test load, designs a set of APU dummy loading device to introduce the load to complete the static test. The finite element analysis model is verified by the test results. The verified finite element analysis, limit load and limit load static strength test are used to show the airworthiness compliance of the static strength of the connection between APU and fuselage.

1. Introduction

APU is a set of self-contained small engines installed on the aircraft that do not rely on any energy outside the aircraft. During ground maintenance of the aircraft, the main engine does not work, and APU can provide power, gas, hydraulic and other energy to the aircraft[1]. APU is generally installed in the tail cone area of the aircraft rear fuselage, which is an independent small turbine power unit[2].

In terms of strength analysis and test of civil aircraft structure, Xintao Wang[3] studied the loading design of the complex space joint of the test piece, realized the effectiveness of loading and ensured the strength of the loading joint itself. Weiping Li et al.[4] developed a set of test system for the static test of the suspension section of civil aircraft engine according to the technical requirements of the static test of the suspension section, and solved the key technologies such as suspension support simulation and loading boundary simulation. Zhiquan He et al.[5] studied the full-scale static test load design technology of the leading edge slats of large civil aircraft, and realized the assessment of the structural safety of the slats and the verification of the strength analysis method. Fafu Zhang[6] studied the application of damage tolerance analysis method in APU installation strength calculation, established the finite element model of important components of the installation system, carried out crack growth analysis and residual strength analysis, obtained its crack growth life, and formulated the inspection method and inspection cycle. Bo Li et al.[7] discussed the general design requirements and design analysis methods of the layout form of the pull rod and its force transmission analysis, vibration isolator design, and installation section design in the design process of the installation system, providing a reference for the schematic design and detailed design of the APU installation system. Zhengyi Zhu[8] studied the full-scale static test verification technology of the composite rear fuselage of large civil aircraft, providing a verification method for the airworthiness verification of the rear fuselage structure. The above scholars' research results on aircraft static test provide research methods and technical parameters for this paper from many aspects, such as load screening, load design, boundary simulation, load introduction mode and so on.

In this paper, the static strength verification of the connection between APU and fuselage is studied, and the stress analysis is carried out according to the connection characteristics of APU and rear fuselage of a civil aircraft. Based on the finite element analysis internal force of APU pull rod of a civil aircraft, the test load screening is carried out, and a set of load screening method for APU fuselage connection test of civil aircraft is proposed. The test results show that the strain data measured by the theoretical analysis and the test are in good agreement. The finite element model is verified by comparing the finite element analysis and the test results. The internal force of the verified finite element model is extracted for the engineering analysis of the connection between APU and fuselage. The verified analysis and static strength test jointly show the compliance with the airworthiness terms.

2. Introduction to auxiliary power unit

APU is installed and fixed on the load-bearing frame of the fuselage by using the mounting rod system with vibration isolator. The APU system consists of three vibration isolators to form three main load-bearing installation joints to realize the load transfer between the APU and the fuselage under normal conditions. The APU installation structure is shown in Figure 1. The number of pull rod is shown in Figure 2.

Figure 1 APU installation diagram

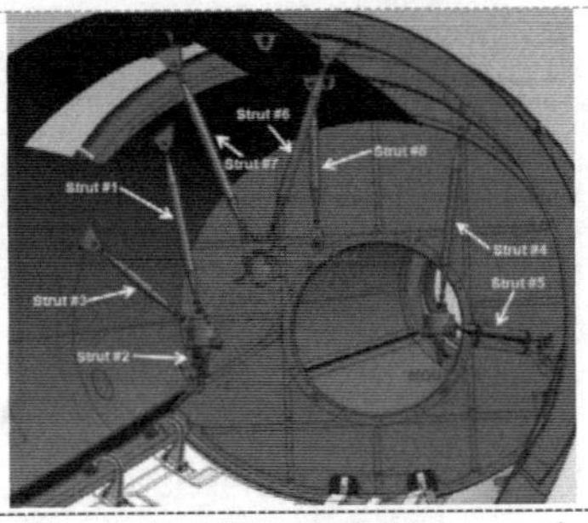

Figure 2 APU installation rod number diagram

3. Static test

3.1. Test support and loading

The test and verification of the connection between APU and fuselage adopts the whole aircraft support mode: constrain the vertical displacement of the front landing gear; constrain the vertical displacement and heading displacement of the left and right main landing gears; constrain the lateral displacement of the left landing gear.

The static test of the connection between APU and rear fuselage adopts APU to load the dummy. The weight of APU is equivalent to that of the dummy, and the center of gravity of the dummy is consistent with that of the real part. It includes two parts: APU body and pull rod system. Of which, the pull rod system adopts real parts, and they are used to connect the false parts of APU body and fuselage structure in the test. APU loading dummy can realize the heading loading, vertical loading and lateral loading.

3.2. Load screening method

The flight load, dynamic load, ground load and gyro load are considered in the static test load of the connection between the rear fuselage of a civil aircraft and APU.

The finite element model of the fuselage rear section structure installed by APU is shown in Figure 3. The finite element model of APU connection is shown in Figure 4. The pull rod load can be calculated by applying the APU overload coefficient to the APU finite element model, and the test load can be screened for the pull rod load. In the APU finite element model, the APU pull rod is simplified into a rod element (rod) and connected with the frame node element, as shown in Figure 4.

The APU body (including lubricating oil and shock absorber), air outlet system and air inlet system are simplified into mass points, and the damper frame is simplified into a beam element (bar). The FEM displacement result is shown in Figure 5.

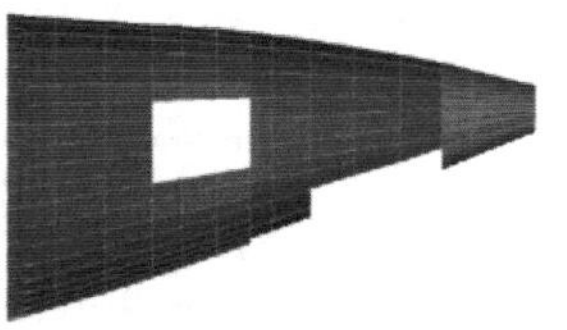

Figure 3 Rear fuselage finite element model

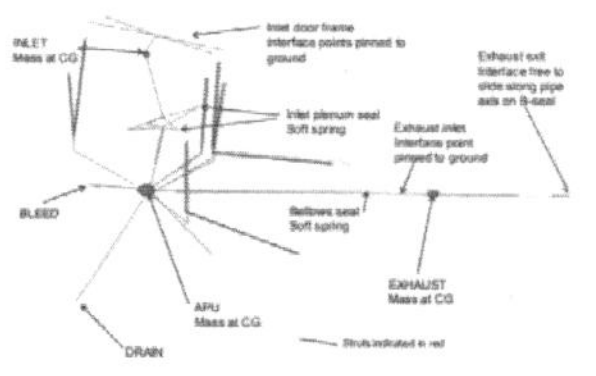

Figure 4 APU finite element model

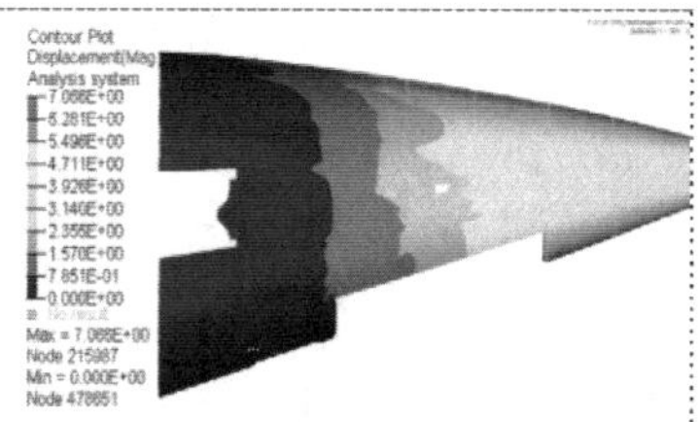

Figure 5 Displacement l result

By calculating the measured value of acceleration environment test of aircraft airborne equipment, this paper obtains the maximum inertial load coefficient (limited load) at the APU body. The inertia loads under different working conditions are loaded into the finite element model of APU installation system, and the load of pull rod (rod unit) can be extracted, as shown in Table 1.

Table 1 Rod load (limit load)

Load case	Pull rod joint load/ (N)						
	Rod 1	Rod 2	Rod 3	Rod 4	Rod 5	Rod 6	Rod 7
Forward 1.00g	370	900	-40	330	900	-260	-140
Backward 0.71g	-263	-852	28	-234	-852	185	99
Down 6.99g	4 893	70	-70	4 683	70	3 425	3 495
Up 3.35g	2 345	-34	34	2 245	-34	1 642	1675
Righ 2.45g	1 936	343	4 655	1397	-294	2 940	2 940
Left 2.45g	1936	-343	-4 655	1397	294	2 940	2 940

3.3. Consideration of gyro load

The gyro load calculation formula is:

$$M = I_x \times \omega_1 \times \omega_2 \tag{1}$$

Where (1): M is the gyroscopic moment generated by the engine; I_x is the moment of inertia of APU rotor; ω_1 is the rotating angular speed of APU rotor; ω_2 is the maneuvering pitch or yaw angular speed of the aircraft.

Through the gyro load calculation, because the rod force generated by the gyro load is very small compared with the rod force generated by the inertial load, the gyro load can not be considered in the test.

3.4. Load screening results

Through the load screening process, three dangerous working conditions under limited load can be obtained: downward 6.99g, forward 1.00g and right 2.45g. The test load is shown in Table 2.

Table 2 APU limited load

Serial number	Load cases	Fx/(N)	Fy/(N)	Fz/(N)
1	Forward 1.00g	-1 925	0	0
2	Right 2.45g	0	0	4 716
3	Downward 6.99g	0	-13 455	0

4. Test result

The forward 1.00g working condition is the limited load working condition, and the assessment structure is tie rod No. 2 and tie rod No. 5. The finite element analysis and calculation strain results and test measured strain of the pull rod under the working condition of 1.00g are shown in Figure 6.

Under 2.45g braking load condition to the right, the assessment structure is tie rod No. 3. The finite element analysis and calculation strain results and test measured strain of the pull rod under the working condition of 2.45g to the right are shown in Figure 7.

The downward 6.99g working condition is the limited load working condition, and the assessment structure is tie rod No. 1 and tie rod No. 4. The finite element calculation strain results and test measured strain of the pull rod under the working condition of 6.99g are shown in Figure 8.

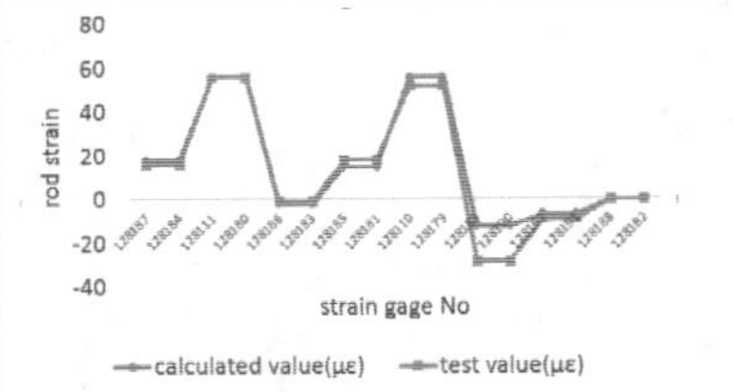

Figure 6 Rod strain comparison (1.00g load case)

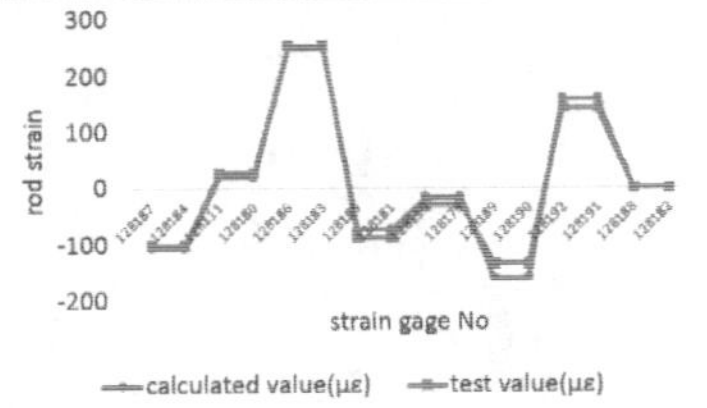

Figure 7 Rod strain comparison (2.45g load case)

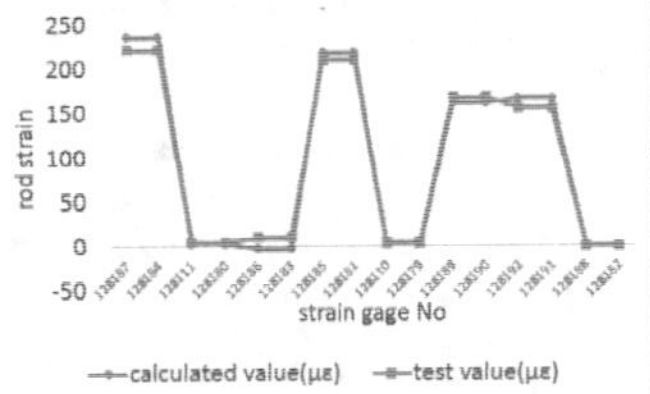

Figure 8 Rod strain comparison (9.00g load case)

5. Conclusion

In this paper, a set of static test scheme for APU fuselage connection of civil aircraft is proposed, and the test load screening method based on APU pull rod load is proposed, and the limited and ultimate load verification conditions are determined.

This paper presents a set of finite element modeling method for the connection between APU and fuselage. Through the comparison between the finite element analysis value and the test measurement value of the connection structure between APU and fuselage, it can be found that the strain comparison error is basically within 5%, indicating that the finite element model modeling is relatively reasonable, and the verified finite element analysis can be used for the subsequent airworthiness compliance verification.

References

[1] Li Dongjie. Application Status and Development Trend of Auxiliary Power Unit [J]. Aeronautical Science & Technology, 2012(6): 7-10.

[2] Zhang Xu, Xiaoning Guo. Static Strength Design of Auxiliary Power Unit Mounting System in Commercial Aircrafts [J]. Science Technology and Engineering, 2011(30): 7476-7480.

[3] Xintao Wang. Design and Analysis of Wing Fixture for Aircraft Structural Strength Test[J]. Engineering & Test , 2019(03): 131-132.

[4] Weiping Li, Wei Tan, Caijun Xue, Hong Nie. Static Test and Computational Analysis for Pylon of Airliner Engine [J]. Journal of Nanjing University of Aeronautics & Astronautics, 2011, 43(6): 732-737.

[5] Zhiquan He, Yang Liu, Zejiang Li. Load design for full scale static test of slat on large civil aircraft[J]. Acta Aeronautica et Astronautica Sinica, 2019, 40(2): 522197.

[6] Fafu Zhang, Cuichao Li, Hao Chen. Damage Tolerance Analysis for APU Mount Components of Civil Aircraft [J]. Modern Manufacturing Technology and Equipment, 2016(7): 68-71.

[7] Bo Li, Honggang Tang, Haitao Yin, Li Tang. APU Mount System Design and Analysis [J]. Civil Aircraft Design and Research, 2010(3): 42-56.

[8] Zhengyi Zhu, Yukang Reng. Full-scale static test and analysis for civil aircraft rear fuselage [J]. Machine Design and Manufacturing Engineering, 2021, 50(5): 70-74.

[9] Civil Aviation Administration of China. Chinese civil aviation regulations: Airworthiness standards transport category airplanes: CCAR-25-R4 [S]. Beijing: Civil Aviation Administration of China, 2011.
[10] General Editorial Board of Aircraft Design Manual. Aircraft design manual (9th volume): Load, strength and stiffness [M]. Beijing: Aviation Industry Press, 2001: 193-670.

Study of COSPAS-SARSAT Beacon Device Used in Chinese Crewed Spaceflight

Xue Ji[1], Ze Hua He[2, 1], Xing Xue Song[3], Jian She Ye[3], Bei Ming Zhang[3]

[1]China Transport Telecommunications and Information Center, Beijing 100011, China

[2] CETC Maritime Electronics Research Institute (Ningbo), Co., Ltd, 315040 Ningbo, China

[3]Beijing Institute of Tracking and Telecommunications Technology, Beijing 100094, China

jixue@cttic.cn

Abstract: Positioning and carrying out search and rescue as soon as possible after a spacecraft is forced to land or crash is a key issue in the field of the Global Human Spaceflight Program. To promptly obtain an emergency location in the event of an unplanned landing, the Search and Rescue Satellite System (COSPAS-SARSAT) has been widely used in Global Human Spaceflight Program. Due to the currently widely used distress beacon landing crash activation, the spacecraft often lost the chance to survive at that time. Therefore, the Search and Rescue Satellite System regards the Emergency Locator Transmitter for In-Flight Distress Tracking (ELT-DT) alike device as the main research direction of technology. This paper introduces the application of COSPAS-SARSAT in global human spaceflight programs, the operation performance of the ELT-DT, the operation of the custom ELT alike device on the re-entry module and the prospects of the application.

1. Introduction

Accurately locating the spacecraft when it lands and quickly conducting the rescue work is an important part of the Global Human Spaceflight Program. Currently, the re-entry module of the crewed spacecraft adopts a dynamic adaptive method. The target landing point is determined before the re-entry module returns. The ground command centre will simultaneously deploy radar positioning systems and rescue forces near the estimated landing point. However, when the re-entry module lands outside the planned route or locates in extreme environments such as deserts and overseas, the Search and Rescue Satellite System (COSPAS-SARSAT) would play a very vital role.

COSPAS-SARSAT was jointly established by the United States, the former Soviet Union, France and Canada in 1979. This system aims to use search and rescue satellites to achieve distress location services on a global scale. It is one of the important components of the International Maritime Organization (IMO) Global Maritime Distress and Safety System (GMDSS). COSPAS-SARSAT adopts two positioning methods, Global Navigation Satellite System (GNSS) positioning and independent positioning which only depends on the beacon alert itself. Hence, even when GNSS fails due to the external environment, then independent positioning can be used for backup [1].

COSPAS-SARSAT is committed to searching and rescuing human life without distinction and discrimination.

Even if some countries are not officially members of COSPAS-SARSAT, they can still obtain distress information through the Search and Rescue Point of Contact (SPOC) with the support of the Mission Control Center (MCC). In 2021, 3623 people from at least 1149 incidents were rescued with the assistance of COSPAS-SARSAT [2]. This system is mainly composed of the ground segment, space segment, beacon and SPOC. The MEOSAR system which consists of 46 medium-orbit satellites makes it possible to locate globally and in real-time, as China also launched six medium-earth orbit BDS satellites that carry search and rescue payloads [3]. Currently, 406 MHz Emergency Locator Transmitters (ELTs), Emergency Position-Indicating Radio Beacons (EPIRBs), and Personal Locator Beacons (PLBs) are commonly used. COSPAS-SARSAT has been widely used in global human spaceflight programs [4].

With years of technological development, the application of COSPAS-SARSAT in the Global Human Spaceflight Program has made great progress. However, at present, the activation of the beacon still adopts the landing crash or manual timing. The distress location is a fixed point in this situation. With the environment more and more complicated, the demands for automatic activation in space and continuous distress position generating are increasing. Hence, the Emergency Locator Transmitter for In-Flight Distress Tracking (ELT-DT) is developed, which can be automatically triggered when sensing abnormal spacecraft behaviours, and generate traces for tracking. The application of ELT-DT will greatly improve rescue efficiency.

This paper introduces the application of COSPAS-SARSAT in global human spaceflight programs, the operation performance of the ELT-DT, the operation of the custom ELT alike device on the re-entry module and the prospects of the application.

2. The application of COSPAS-SARSAT in the global human spaceflight program

Unlike the regular beacon mentioned before, as a special beacon, the custom beacon is used in most global human spaceflight missions.

2.1. *Russia*

The former Soviet Union has sent Gagarin, the first human astronaut, into space since the 1960s. It is a veteran aerospace power in the world. Currently, the Russian re-entry module uses a custom ELT for positioning.

2.2. *The USA*

The National Aeronautics and Space Administration (NASA) is installing beacons for its Artemis program Orion spacecraft research. The Artemis program is a U.S government-funded human spaceflight program that aims to send astronauts to the moon and back safely by 2024 [5]. Establish a normalized residency mechanism to pave the way for future manned landing missions on Mars. The Orion spacecraft is a new generation of crewed spacecraft developed and is the main carrier of the U.S Mars manned landing program.

2.3. *China*

Since the development of the Shenzhou-1 mission in 1998, COSPAS-SARSAT has participated in all the Manned series spaceflight programs.

Consider wrong orbital maneuvers, wrong trajectory detection, space collisions, and re-entry capsule landing location is possibly out of the radar coverage area. Hence, COSPAS-SARSAT is used in emergency operation. It provides timely and accurate landing point information for the emergency return and normal recovery of the Manned series spacecraft, which enhances the efficiency of finding the module, and decreases coordination tasks for foreign authorities.

3. System level test of ELT-DT

3.1. The working mechanism of ELT-DT

When distress occurs, ELT locates a single point position, but ELT-DT will generate a large number of continuous position messages. During 0 to 2 minutes of ELT-DT activation, MCC gets a location every 5 seconds and forwards each position to Rescue Coordination Center (RCC). Then MCC gets a location every 10 seconds for 2 to 5 minutes, and every 28.5 seconds during beacon activation for 5 to 15 minutes. This means there are about 60 total messages in the first 15 minutes. Next, further location information is updated every 10 minutes after 15 minutes of activation.

Since ELT-DT can generate continuously updating positions during movement, the running trajectory of the spacecraft is formed from points to lines, which improves the ability to detect abnormal flight and replay the state afterward [6].

The specification difference between the ELT and ELT-DT is shown in Table 1[7].

Table 1. The specification of the ELT and ELT-DT

	ELT	*ELT-DT*
Protocol	10	1
Repetition Period	50 seconds	5/10/28.5 seconds
Cancellation message	NO	YES
Medium-term stability	YES	NO
Output rise time	less than 5 MS	less than 2 MS
Transmitter Power	35 to 39 dBm	36 to 39 dBm
The minimum operation time	24 hours	370 minutes
Beacon Activation	within 15 sec	within 5 sec
Antenna Characteristic	3 d Bi and4 d Bi over 90%	2 d Bi and 6 d Bi over 90%

3.2 Testing for ELT-DT in China

COSPAS-SARSAT conducted the Do-No-Harm for ELT-DT test in January 2022. The main purpose of the system test is to verify the performance of new components and related functions by expectations in the entire system through end-to-end full link tests on related equipment. Further, the Beijing MEOSAR local user terminal (BEIJING-MEOLUT) in China has already been commissioned for the function of ELT-DT in April 2021. CNMCC also participated in this global ELT-DT test.

Figure 1 shows the geographic locations of Turkey and China. It can be seen from the figure that China is far away from Turkey, almost spanning two hemispheres.

Figure 1. The geographic location of Turkey and China.

One beacon simulator in Turkey ran the ELT-DT script on January 12th and 19th. The Turkish beacon simulator performed a long-range flight from Misrata, Libya to Ankara, starting from the ALMCC region (Algeria) and ending in the TRMCC region (Turkey), passing through the ITMCC (Italy) and

GRMCC (Greece) regions respectively. The cross-regional design of the script is to detect whether the MCC data distribution is normal in a complex environment. The beacon ID of the Turkey simulator is 21F25787C4BFDFF.

Due to the abnormal signal of the beacon simulator on the 12th, the technical expert group did not consider the statistical analysis of this data from MCC in global. The beacon simulator totally sent 326 bursts on the 19th. BEIJING-MEOLUT received 449 packets which are relayed from different satellites. There is a GNSS position in each packet. The system will detect the time and frequencies when packets transmit and when BEIJING-MEOLUT receives. Using these data can solve the joint equations which generate 43 independent positions in total.

On the 19th, the first independent position was generated 42 seconds after BEIJING-MEOLUT detected the first Turkish simulator's beacon packet. Table 2 shows the processing performance of MEOLUT for ELT-DT. The GNSS positioning data of the beacon which is written in the data packet means that there is a GNSS position for each data packet received.

Table 2. The processing performance of BEIJING-MEOLUT for ELT-DT in Turkey.

Date	*Packet/GNSS position*	*Independent position*
12th	298	32
19th	449	43

The 12th is a test drill so the beacon simulator emits less semaphore. The system test results in January 2022 show that the ELT-DT can be detected, received, decoded and forwarded to the corresponding responsible unit by the COSPAS-SARSAT system in a timely and highly consistent manner.

4. Performance of the beacon in the recent spaceflight Mission

4.1 Preparation

Since the new-generation crewed spacecraft test chamber carried by the Long-match 5B launched successfully, Chinese MEOSAR ground facilities have also begun to play a role [8]. Various performance indicators have improved significantly: the positioning time shortened from 40 minutes to 3 hours in the past to less than 1 minute. The position error is shortened to 500 meters.

The spacecraft normally consists of a service module, a re-entry module, and an orbital module. The custom beacon that has 2 sets of antennas is installed in the re-entry module. Normally, when the re-entry module is approaching the landing, it will be triggered according to the program settings. Then the beacon will transmit a burst around a mean value of 50 seconds. LUT generates the position after receiving and processing the burst. Next, the CNMCC reports the position to the ground command center in time.

To ensure the high-quality completion of the support mission, CNMCC has participated in many seminars on the design of the landing site. CNMCC, together with relevant departments, has carried out beacon tests for several scenarios such as global mode, local mode, air-ground coordination and search and rescue coordination. This work provides a solid foundation for formal mission assurance.

4.2 On orbit test

The astronauts also conducted a test at 15:25 UTC, 15th April to locate the re-entry module, as the module still running in regular orbit. Although the test lasted only a short time, two locations were determined which could be seen in Figure 2. There are three MEOLUTs: Taupo, Toulouse and Beijing receiving the two-dimensional positioning information.

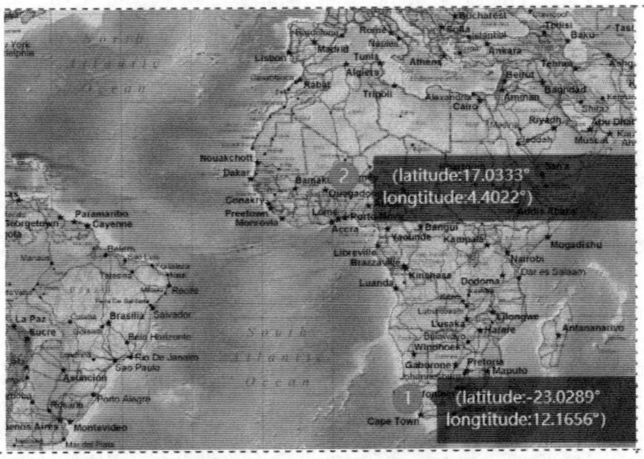

Figure 2. The astronauts test in space.

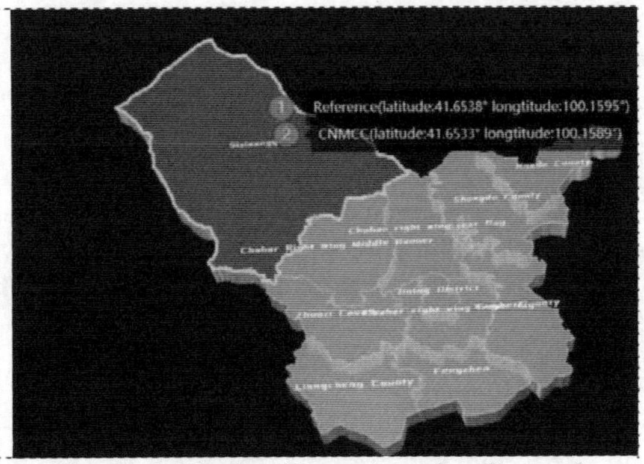

Figure 3. The landing point of reality and COSPAS-SARSAT.

4.3 Landing tracking.

At 9:56:12 on June 16th, the re-entry module successfully landed in the predetermined area of the Dongfeng Landing Site. After the spacecraft passed the "black barrier", the beacon was activated and sent signals during the parachute opening phase. CNMCC accurately located the landing point of the re-entry module. The reference position is located by ground radar. The positioning accuracy between the reference and GNSS is around 100 meters, and 200 meters between the reference and independent position, as can be seen in Figure 3, Figure 4 and Figure 5.

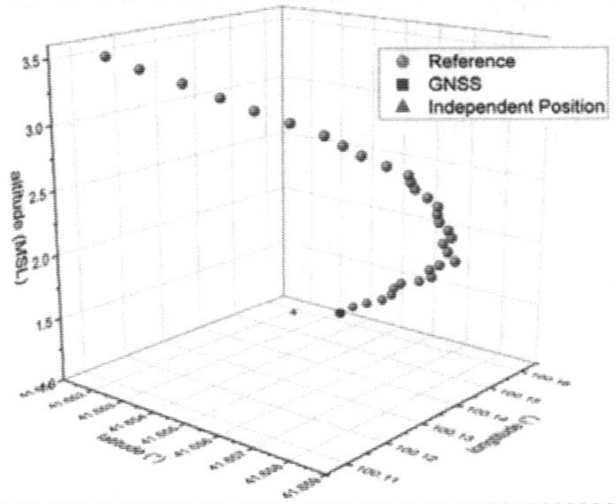

Figure 4. The landing point of reality and COSPAS-SARSAT.

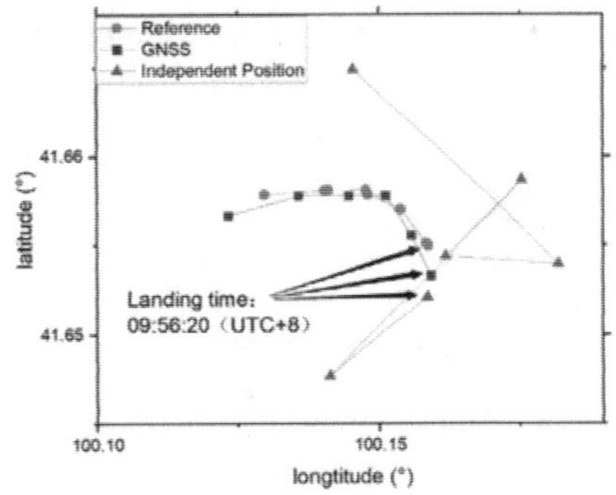

Figure 5. The error of reality and COSPAS-SARSAT.

4.4 Conclusion

Based on the above data, the current GNSS and independent positioning accuracy are within 200 meters in the spaceflight Mission. When the re-entry module encounters an unexpected forced landing, the beacon can be activated immediately and COSPAS-SARSAT will generate accurate positions for rescue. According to the result, implemention of COSPAS-SARSAT system improved the timeliness of positioning and the effectiveness of the coordination of rescue forces. Various customized beacons are being developed to adapt to various scenarios during the return of manned spaceflight.

At present, COSPAS-SARSAT ground facilities cannot obtain the trajectory before the black-barrier. Because COSPAS-SARSAT is not designed for aerospace missions. However, the ELT-DT alike device can be automatically triggered and activated when the velocity, acceleration or any other characteristic are changed in the space. Furthermore, its special processing mechanism can realize continuous three-dimensional position generating of the re-entry module with altitude besides Latitude and Longitude.

Meanwhile, CNMCC estimated the route of the re-entry module with the help of the continuously transmitted messages of the ELT-DT.

5. Future prospects

The use of ELT-DT in the Global Human Spaceflight Programs will greatly improve the timeliness of locating after distress to improve the survival rate. The research on ELT-DT of COSPAS-SARSAT has entered an advanced stage and is expected to provide formal services for global before January 1, 2023. Also, it is expected to reach within 100 meters in the future combined with the second-generation beacon application [9].

Acknowledgments

This work was supported by the National Key Technologies R&D Program of China (Grant No. 2021YFB3901503).

References

[1] COSPAS-SARSAT. System introduction http://cospas-sarsat.int/en/system-overview/cospas-sarsat-system
[2] COSPAS-SARSAT Secretariat. Summary of Draft COSPAS-SARSAT Report on System Status and Operations No. 38 (January - December 2021).
[3] G. Li, S. Guo, Z. He, Y. Gao, W. Li. BDS-3 SAR Service and Initial Performance. GPS Solutions. (*GPS Solutions* (2021)) 25:134.
[4] W. Sun, Y. Hou. Application of Global Satellite Search and Rescue System in Manned Space Mission Search and Recovery. (*Manned Space Flight*, 2003), 03(No.3):62-66.
[5] NASA. Artemis plan https://www.nasa.gov/press-release/nasa-publishes-artemis-plan-to-land-first-woman-next-man-on-moon-in-2024
[6] Q. Wan, Y. Chen, X. Ji. The Study of the Application of the Emergency Locator Transmitter for In-Flight Distress Tracking for China Civil Aviation Safety. (China: *ICCASIT* (2021)) pp 111-114.
[7] COSPAS-SARSAT. "Specification for COSPAS-SARSAT 406Mhz distress beacons". C/S T.001
[8] X. Ji, Y. Chen, Q. Wan. Design and Research of MEOSAR Ground Receiving Station in COSPAS-SARSAT. *Modern Radar* (2021)11
[9] COSPAS-SARSAT. "COSPAS-SARSAT MEOLUT Performance Specification and Design Guidelines". C/S T.019.

Station-keeping of Stratospheric Aerostat with Receding Horizon Tree Search Planning Method

Mengshan Xie[1], Yuan Long[1], Xiaolong Deng[1,*], Xixiang Yang[1]

[1] College of Aerospace Science and Engineering, National University of Defense Technology, Changsha, Hunan, China

[*]xiaolong.deng@outlook.com

Abstract. The station-keeping property of stratospheric aerostat is significant for practical applications. In this work, we established the dynamics model of aerostat, then divided the wind fields into various layers according to the wind direction variation, which converted the data that wind speed continuously varies with height into a set of discrete wind layer sequences. After that, in order to realize station-keeping, the receding horizon tree search strategy was implemented to select the candidate wind layers. Finally, the station-keeping trajectory of the aerostat in dynamic wind fields was simulated. The results have shown that the proposed method could make the selection of the perpetual altitude more reasonable, and greatly reduce the computational complexity, effectively achieving the purpose of keeping the aerostat in a certain area.

1. Introduction

Stratospheric aerostat is a type of important platform for research and application in near-space, which has significant advantages of large load capacity, high flight altitude, and high cost-effectiveness in the fields of remote sensing, scientific research, and communication [1,2].

Researchers have proposed a type of novel stratospheric aerostat system based on the use of the wind field environment, with the idea of rationally utilizing the wind speed and direction at different altitudes [3]. Typically, the aerostat can change altitude through auxiliary components, such as a valve or ballonet [4]. By adjusting altitudes, aerostat can realize missions of station-keeping or trajectory planning [5].

Saito et al. [6] proposed a tandem balloon system with long-endurance stationary flight capabilities. The system consists of a zero-pressure main balloon in series with an over-pressure balloon located below the load. Through a retractable suspension rope, the station-keeping performance of the tandem balloon system has improved, changing the altitude of the main zero-pressure balloon with a little energy. Therefore, the altitude changes of a traditional zero-pressure balloon due to the temperature difference can be controlled, and the wind layers with different wind directions can be used for trajectory control.

Born et al. [7] indirectly controlled the trajectory in the horizontal direction of the aerostat through altitude control and adopted a method of stratification of altitude to reduce the candidate altitudes. Finally, they calculated the local optimal trajectory through the search algorithm.

In this paper, a receding horizon tree search algorithm is implemented to analyse the station-keeping trajectory of an aerostat in a dynamic wind field, which verifies the effectiveness of the optimization algorithm for the station-keeping of the aerostat.

2. Stratospheric aerostat dynamics model

As shown in Figure 1, stratospheric aerostat realizes station-keeping based on wind field utilization through altitude control [8]. During the process, a stratospheric aerostat is mainly subjected to three forces: the buoyancy provided by the lift gas in the capsule, the gravity and the dynamic drag. Since the aerostat studied in this paper does not have horizontal control capabilities, the movement of the aerostat in the horizontal direction entirely depends on the wind [9].

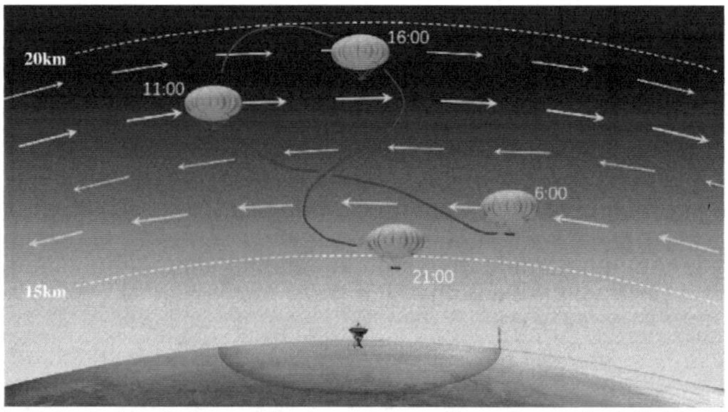

Figure 1. Principle of stratospheric aerostats based on wind field utilization

The agent is assumed to exhibit Lagrangian dynamics, that is, the horizontal velocity $\dot{x}(t)$ and $\dot{y}(t)$ are equal to the wind field velocity, and the vertical velocity is equal to the wind field velocity summed with the control effort. Meanwhile, Gaussian white noise is added in both the horizontal and vertical directions. The nonlinear dynamics are described by [10]

$$\begin{aligned} \dot{x}(t) &= f_x\left(x(t), y(t), z(t), t\right) + w_x \\ \dot{y}(t) &= f_y\left(x(t), y(t), z(t), t\right) + w_y \\ \dot{z}(t) &= f_z\left(x(t), y(t), z(t), t\right) + u(t) + w_z \end{aligned} \tag{1}$$

Where f_x, f_y, f_z is the wind speed in the x, y, and z directions at the position $((x(t), y(t), z(t))$ of the aerostat at the current moment, w_x, w_y, w_z are wind field noises in the three directions with standard normal distribution, $u(t)$ is the control effort of the aerostat in the vertical direction.

When x, y, and t are sufficiently small, it can be assumed that $f_x\left(x(t), y(t), z(t), t\right)$ in Equation (1) is only a function of z; and stratospheric wind velocity in the vertical direction is negligible since it is generally small compared to the control effort of the aerostat. In addition, we do not take wind field noise into account while planning. Those assumptions simplify Equation (1) to

$$\begin{aligned} \dot{x}(t) &= f_x\left(z(t)\right) \\ \dot{y}(t) &= f_y\left(z(t)\right) \\ \dot{z}(t) &= u(t) \end{aligned} \tag{2}$$

3. Wind stratification method

A certain altitude range (15~25km selected) is analysed according to the wind properties [8]. And we divided the range according to the wind direction gradient.

Take the initial altitude as 15km, gradually increase the thickness of the wind layer with a resolution of 10 meters from bottom to top and calculate the variance of the wind direction value at all altitudes in the wind layer.

When the variance exceeds a threshold (5 in this process, which is related to the altitude resolution), the increase of the thickness is stopped. At this point, if the thickness of the wind layer surpasses a threshold (500m and 50m), the layer is considered a candidate wind layer, otherwise, it is discarded.

Applying this stratification method, we made some attempts.

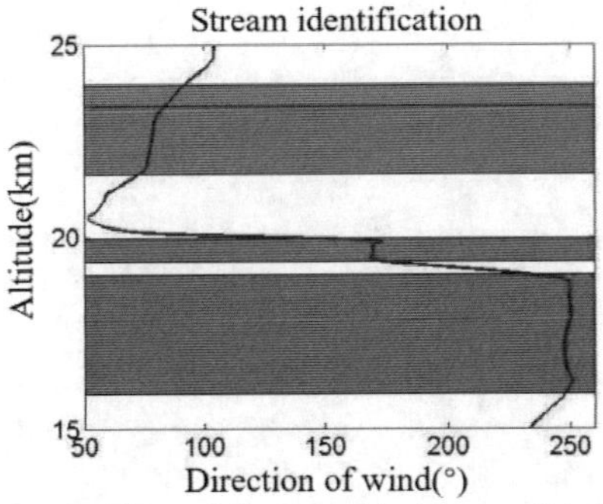

Figure 2. The minimum wind layer thickness of $500m$ and maximum variance of 5

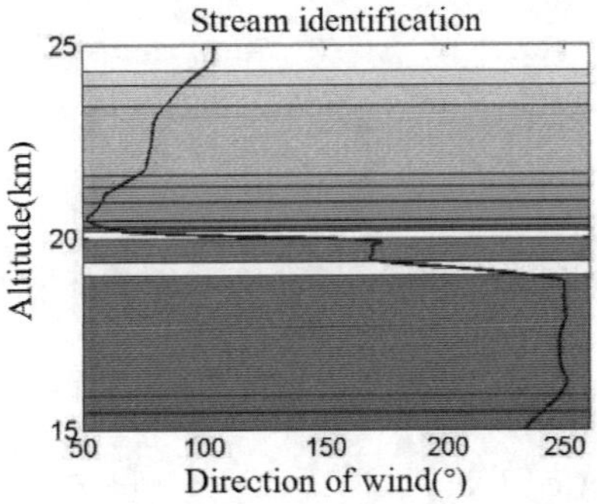

Figure 3. The minimum wind layer thickness of $50m$ and maximum variance of 5

4. Receding horizon tree search model

The tree search method is a process of listing all the available options for each step and extending them like the roots of a tree. In this paper, each optimization step is a transfer of wind layers and the number of branches for the next optimization stage is determined by the number of wind layers available.

Assuming that the area in which the stratospheric aerostat works is divided vertically into three wind layers, thus the aerostat has three options from the initial wind layer: move to the other two wind layers or stay at the original one. When the next step is reached, the aerostat has three choices again, and so on, to find all the possible choices for a given number of steps.

Considering the computational complexity, we used a "branch pruning" method where after each step, the incurred cost is calculated and then compared with the current optimal cost. If it is greater than the current optimal cost, this branch has no further action, instead, other branches are explored.

Since the aerostat altitude needs to be adjusted several times during the station-keeping and the fact that the wind speed changes with altitude in a dynamic wind field. We use an optimization method in receding horizon control to dynamically select the executive operations of the aerostat. This optimization strategy is summarized as follows:

(1) Solve the optimal control problem during a fixed future period $[i, i+N-1]$ under the current state x_i, with both current and future constraints;

(2) Execute the first step only in the optimal control sequence $\{u_i^*, \cdots, u_{i+N-1}^*\}$ obtained and measure the state at the next moment $i+1$ after the control;

(3) Repeat step (1) now $i+1$ to solve the problem for the state x_{i+1} over the future period $[i+1, i+N]$.

In the above, it is assumed that the state was measured at moment $i+1$ and we can measure the state x_{i+1} in practicality through some forms of observations based on the available data, so the hypotheses that all states have been measured and noise is ignored are reasonable.

After the optimization process, the nodes selected by the receding horizon tree search represent the wind layers filtered, and the branches are the forward-simulated dynamics between nodes. The branches differ in terms of duration and physical distance since different layers are in various locations.

During the process, the total number of branches in the decision tree is Cd, where C is the number of wind layers, d is the search depth, referring to the number of wind layers transferred.

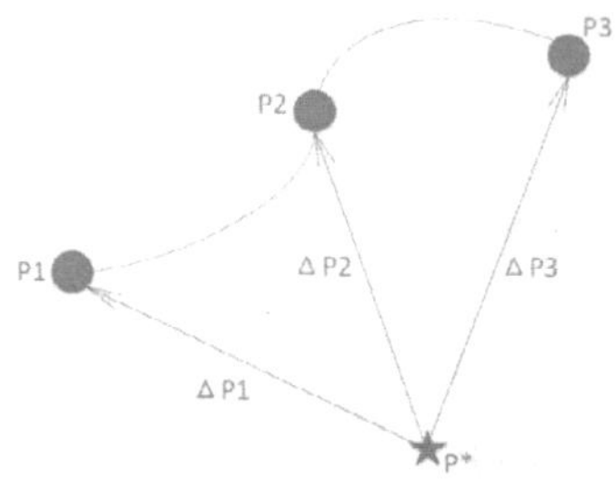

Figure 4. Calculation of the cost function in the horizontal direction

We set each branch cost Δp_i as the distance between the current wind layer and the target point (x^*, y^*) in the horizontal direction:

$$\Delta p_i = \left(\left(x_i - x^* \right)^2 + \left(y_i - y^* \right)^2 \right)^{1/2} \tag{3}$$

And the total cost is $J = \Delta p_1 + \cdots + \Delta p_d$. Applying Equation (3), we can get the specific cost.

A brand can join a node to itself, which means the aerostat choose to stay in it rather than move to other wind layers. In this case, the position of the aerostat about the target needs to be determined. If the aerostat tends to 'move away' from the target, it is controlled to stay in the current wind layer for a transient period (3 minutes in this case) and then leave, ending the search for this branch. If the aerostat is moving towards the target, it is asked to stay in the current wind layer until the situation is reckoned to have changed sufficiently (e.g., a change in the speed sign of the aerostat relative to the target), ending the planning.

To get the station-keeping, the objective is to satisfy the following equation in such a way that the total cost function J_p is minimized in the process of changing the wind layer altitude z_k [11]:

$$\begin{aligned} \underset{z}{\text{minimize}} \quad & \sum_{k=0}^{T} J_p \\ \text{subject to} \quad & x_{k+1} = x_k + f_x(z_k)\Delta t, \\ & y_{k+1} = y_k + f_y(z_k)\Delta t, \\ & z_{k+1} = z_k + u_k \Delta t, \\ & z_{\min} \le z_k \le z_{\max}, \forall k. \end{aligned} \tag{4}$$

5. Simulation results

The parameters of the aerostat in this work are designed according to that of Google Balloon [8]. Since the process of ascending and descending is respectively realized through the exhaust valve discharging the air from the balloonet and the compressor pressing the outside air into the balloonet, we simplified the exhaust valve and compressor model by assuming that the vertical motion is uniform, setting the speed in the vertical direction to a constant value [12]. The key parameters are listed in the Table 1.

Table 1. Key Parameter in this simulation process

Parameter	Value
Initial Altitude	15km
Volume	2500m^3
System Mass	188.84kg
Helium Mass	31.25kg

Pressure Difference	$150Pa$
Vertical Velocity	$5m/s$

The wind field is divided where the aerostat is located into several wind layers with the wind stratification method, which are six continuously varying wind fields with 12-hour intervals.

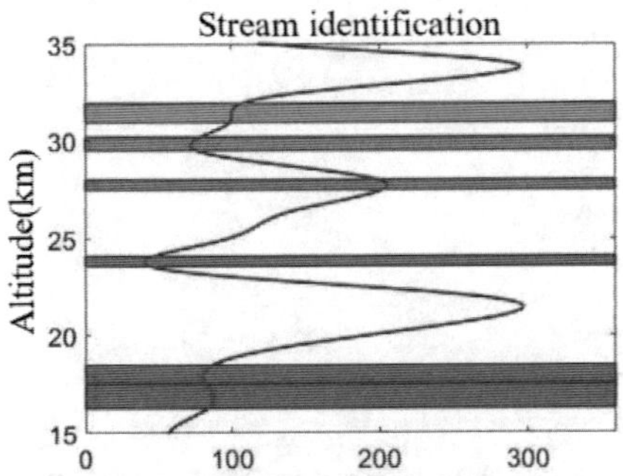

Figure 5. First wind layer division

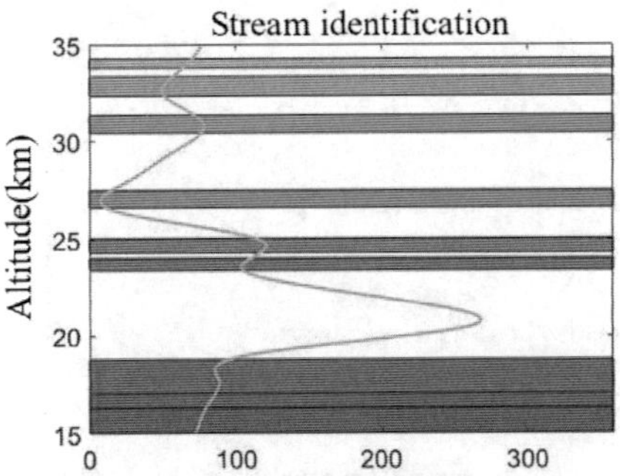

Figure 6. Second wind layer division

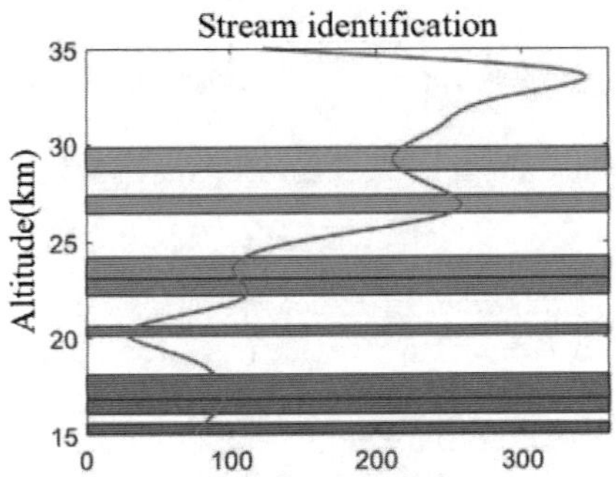

Figure 7. Third wind layer division

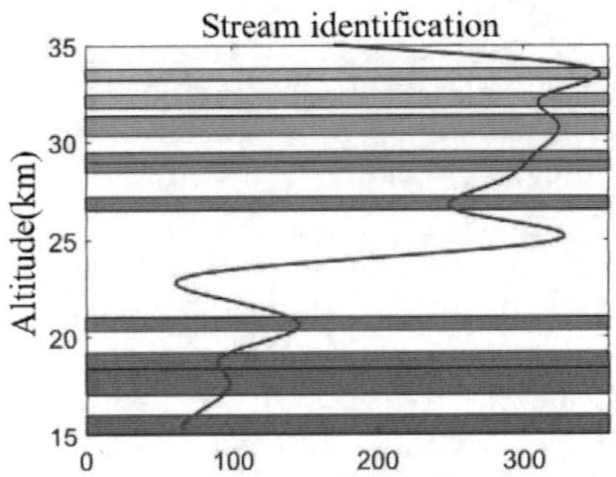

Figure 8. Fourth wind layer division

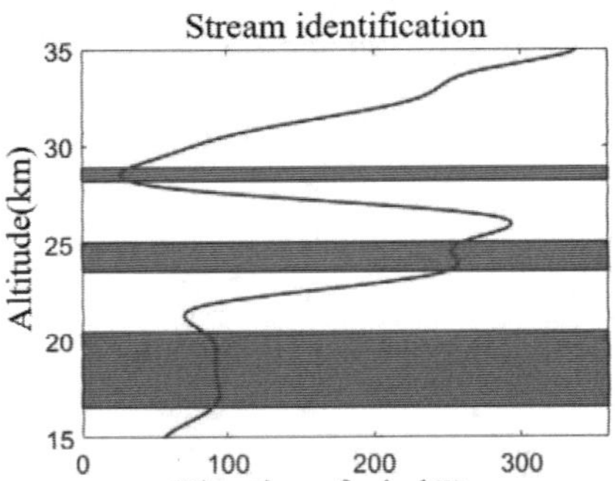

Figure 9. Fifth wind layer division

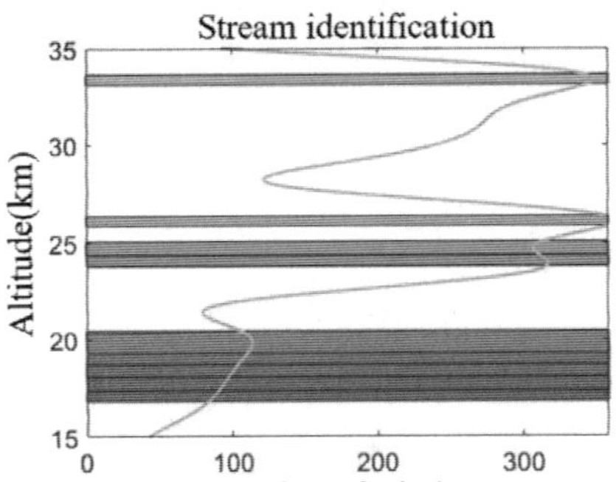

Figure 10. Sixth wind layer division

Taking the duration of wind layer transfer as 4 hours, the aerostat can transfer 3 times in a wind field, which reveals that the depth of the receding horizon tree search is 3, resulting in 1.59×10^{15} possible selection strategies within the 6 continuous wind fields. This is a huge amount of computation. In such cases, we adopt the 'branch pruning' to reduce the computational complexity, getting the simulation results faster.

The results show that the tree search with a depth of 3 performs well and the station-keeping trajectory obtained is shown below.

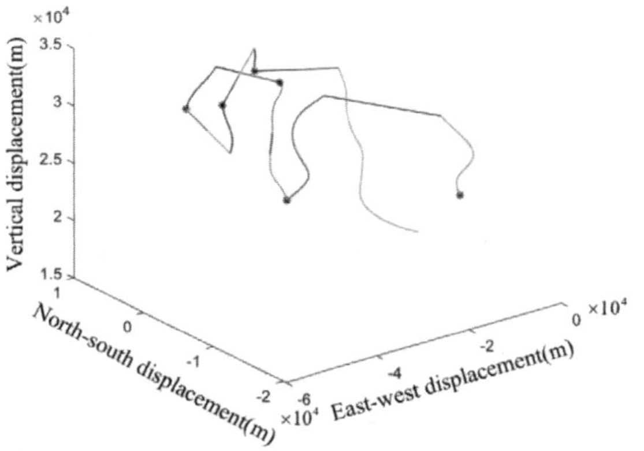

Figure 11. The stratospheric aerostat 3D station-keeping trajectory under the receding horizon tree search strategy

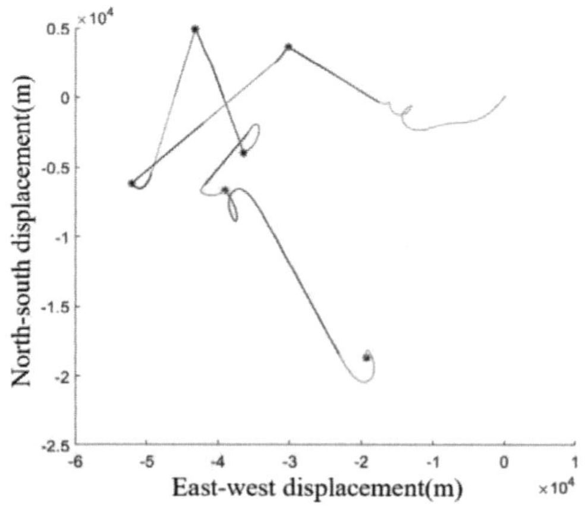

Figure 12. The stratospheric aerostat station-keeping trajectory under the receding horizon tree search strategy in the projection of the X-Y plane

Figure 11 presents the 3D station-keeping trajectory of the stratospheric aerostat using the receding horizon tree search strategy. In the Figure, each curve with a different colour represents a wind layer transition. And the projection of the stratospheric aerostat trajectory in the horizontal plane is shown in Figure 12.

The displacement of the aerostat is about 50km in the east-west direction and 25km in the north-south direction, there is an evident stationing process in the west of the outset. However, as aerostat utilized based on the environment has no control in the horizontal direction, the station-keeping process depends on the environment. If the environment is suitable, the aerostat can implement long-term station-keeping. On contrary, in an unfavorable condition, station-keeping is a tough challenge.

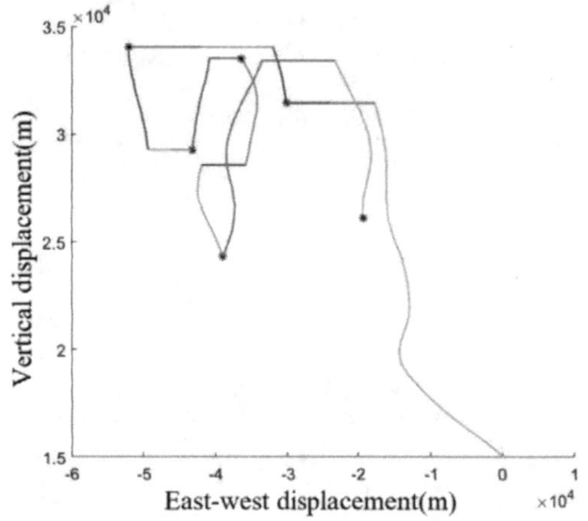

Figure 13. The stratospheric aerostat station-keeping trajectory under the receding horizon tree search strategy in the projection of the X-Z plane

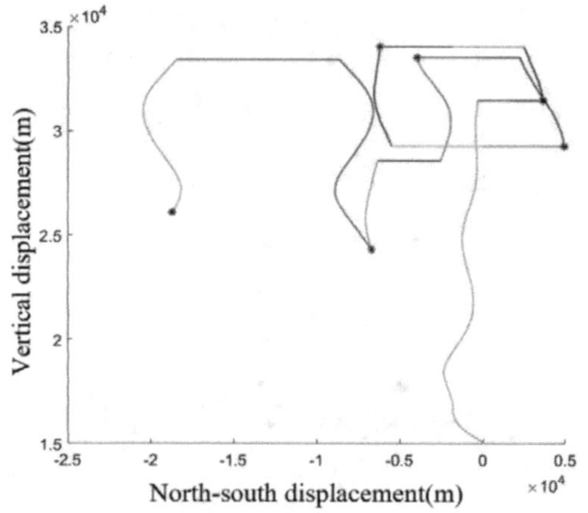

Figure 14. The stratospheric aerostat station-keeping trajectory under the receding horizon tree search strategy in the projection of the Y-Z plane

According to the projection of the trajectory, the aerostat flew westward with the wind field and then flew back and forth through altitude adjustment manoeuvre to achieve station-keeping. And it also remained stationary by adjusting its altitude in the north-south direction. But the aerostat inevitably heads the south due to the wind field, as shown in Figure 14.

6. Conclusion

In this paper, the forces on the aerostat during the station-keeping process are analysed and the dynamics model of stratospheric aerostat is developed. Then wind stratification method is implemented to divide the wind field into several discrete wind layers taking the variance of wind direction values as a reference indicator, effectively avoiding areas that are not conducive to aerostat stationing such as violent wind direction changes, choosing altitude available for aerostat stationing more reasonable.

After that, to achieve station-keeping, the basic principle of the receding horizon tree search algorithm is introduced, and it is used to select the candidate wind layers, finding the optimal wind layer sequence.

Finally, the trajectory of a stratospheric aerostat in a dynamic wind field is simulated and analysed according to this optimal wind layer sequence. The results indicate that the receding horizon tree search algorithm can be used to select the aerostat stationing altitude and effectively achieve the purpose of station-keeping.

Acknowledgments

This work was supported by the National Natural Science Foundation of China [Grant numbers 61903369].

References

[1] Hou Z X, Yang X X and Qiao K. (2019) *Stratospheric Airship Technology* (Beijing: Science Press)
[2] Yajima N, Izutsu N and Imamura T. (2009) *Scientific Balloon* (London: Springer)
[3] DENG X L, YANG X X and MA Z Y. (2019) Review of important technologies for station-keeping of stratospheric aerostats based on wind field utilization. *Acta Aeronautica et Astronautica Sinica*. 40: 022941
[4] Sudha R S, Ma J and Lim K M. (2018) Numerical evaluation of station-keeping strategies for stratospheric balloons. *Aerospace Science and Technology*. 80: 288–300
[5] Zhang J M, Lu Y P and Liu J N. (2007) Research on Trajectory Control Systems for Stratospheric Long Duration Balloons. *Aircraft Design*. 27: 21-24
[6] Saito Y, Akita D, Fuke H and Izutsu N. (2010) Properties of tandem balloons connected by extendable suspension wires. *Advances in Space Research*. 45: 482- 489
[7] Robert D B and Mac S. (2019) Riding an Uncertain Wind Field: Receding Horizon Tree Search Planning with Opportunistic Sampling for an Autonomous Weather Balloon. AIAA SciTech Forum.
[8] Bellemare M G, Candido S and Castro P G. (2020) Autonomous navigation of stratospheric balloons using reinforcement learning. *Nature*. 588: 77-82
[9] Du H F, Lv M Y and Zhang L C. (2019) Energy management strategy design and station-keeping strategy optimization for high altitude balloon with altitude control system. *Aerospace Science and Technology*. 93:105342.
[10] Ramesh S S, Lim K M and Lee H P. (2017) Reduced-order modelling of stratospheric winds and its application in high-altitude balloon trajectory simulations. J*ournal of Applied Meteorology and Climatology*. 56:1753-1766.
[11] Hall J L, Cameron J and Pauken M. (2019) Altitude-Controlled Light Gas Balloons for Venus and Titan Exploration. AIAA Aviation 2019 Forum.
[12] Yang X X, Yang X W and Deng X L. (2020) Horizontal trajectory control of stratospheric airships in wind field using Q-learning algorithm. *Aerospace Science and Technology*. 106: 106100.

Research and Verification on the Solution of Bleed Air Temperature Deviation of Civil Aircraft Pneumatic System

Danqi Peng

Shanghai Aircraft Design and Research Institute, Shanghai 201210, China

pengdanqi@comac.cc

Abstract—The pneumatic system of civil aircraft is equipped with a precooler (PCE) to cool the high-temperature bleed air from the engine to meet the design requirement of a downstream user system. The PCE is usually a fork-flow heat exchanger. This heat exchange structure will make the outlet temperature difference especially large, which shall lead to the deviation of the inlet air temperature of the downstream wing anti-icing system and air conditioning system and cause the temperature of bleed air supplied to the wing anti-ice system unstable. During a flight test of a certain large passenger aircraft, it was found that the bleed temperature at the PCE outlet was lower than the piccolo inlet temperature of the wing anti-ice system when the wing anti-ice system was operated with two engine bleed systems on. This paper creatively proposes design ideas for the three kinds of mixers and determines the solution of adding a mixer downstream of the PCE to solve the temperature deviation problem. The mixer is installed at the hot side outlet of the PCE and used to mix the temperature field at the PCE outlet so that the air supplied to the downstream customer systems meets the design temperature requirements. According to the simulation calculation and laboratory test verification results, the design scheme of a certain large passenger aircraft is selected, which has successfully solved the problem of temperature stratification at the outlet of the PCE and supported the follow-up flight test.

1. Introduction

The pneumatic system of the civil aircraft bleeds air from the engine, Auxiliary Power Unit(APU), or high-pressure ground air source to supply air for the air conditioning system, wing anti-icing system, engine starting system, fuel tank inerting system, and water tank pressurization system, meeting the pressure, temperature and flow requirements of the downstream consumer systems[1].

Since the air bled from the engine is of high temperature and high pressure, it needs to be cooled, which is usually performed by a precooler(PCE)[2-3]. The PCE uses the engine fan air as the cold source, and the fan air flows into the PCE through the fan air valve (FAV) for heat exchange with the engine air. Usually, a bleed air temperature sensor (BTS) is set to measure the outlet temperature of the PCE and transmits the signal to a system controller. After the operation of the controller, the output signal can adjust the opening of FAV and change the fan airflow, so that the outlet temperature of the precooler can be adjusted to a suitable range.

The PCE is usually a fork-flow heat exchanger. The closer the outlet is to the outlet of cold air, the higher the temperature is[4], and the sectional temperature difference at the outlet exceeds 200°C, as shown in Figure 1. Precooler is usually installed on Pylon. After the bleed air is cooled from the precooler, part of it is supplied to the wing anti-icing system located at the wing's leading edge and part to the air conditioning system located at the fuselage. The wing anti-ice air supply line is closer to the PCE outlet, as shown in Figure 2, and because of that, usually BTS is installed in the branch to the air conditioning system. The uneven outlet temperature of the PCE and BTS installation will lead to a deviation of the inlet air temperature of the downstream wing anti-icing system and air conditioning system which shall cause the temperature of bleed air supplied to the wing anti-ice system unstable. Especially will it affect the performance of the wing anti-icing system, leading to the overtemperature of the slat structure and reducing the safety of the aircraft.

During a flight test of a certain large passenger aircraft, it was found that the BTS of the pneumatic system was lower than the piccolo inlet temperature of the wing anti-ice system when the wing anti-ice system was operated with two engine bleed systems on, while BTS was higher than the piccolo inlet temperature with one engine bleed system on.

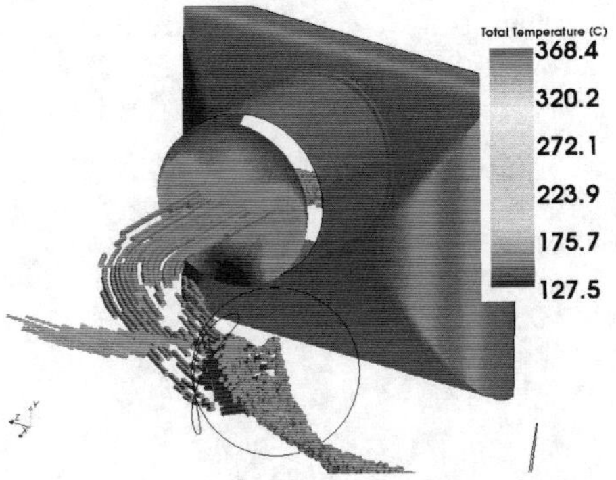

Figure.1 Outlet section temperature of PCE

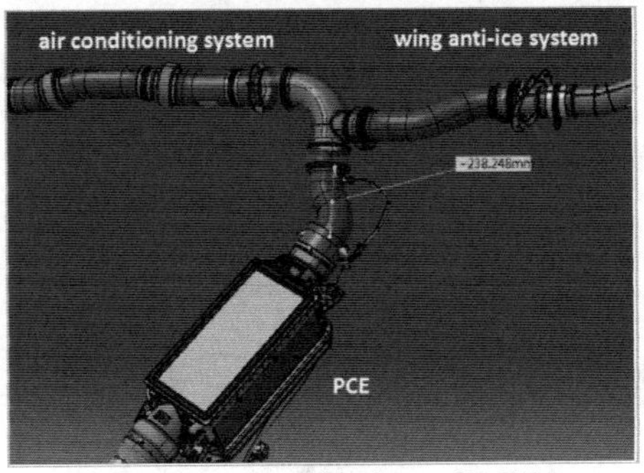

Figure.2 Installation of PCE

To solve the temperature stratification at the outlet of the PCE and the resulting air inlet temperature deviation of the wing anti-icing system and the air conditioning system to make sure that the temperature of bleed air supplied to the wing anti-ice system is within the design requirement, the solutions are as follows: a) Confirm the rule and adjust the set point of temperature control, confirm the regularity of the difference between the BTS temperature and the wing anti-icing inlet temperature through a flight test and laboratory test, and adjust the set point of BTS temperature control through a software modification; b) The precooler or wing anti-ice air supply fork is mixed to weaken temperature stratification; c) Reduce the wing anti-ice air supply flow or temperature.

Considering the change feasibility, change effort, change cycle, and their impact on the system performance, this paper proposes a scheme of adding a mixer at the outlet of the PCE. As the bleed passes through the mixer, it is rotated to achieve temperature uniformity. Considering that the space is small, and the flow resistance should be reduced as much as possible to reduce the impact on the user system. This paper creatively proposes three solutions: 1) Upper and lower flapper mixer; 2) Twin-cyclone mixer and 3) Louver mixer. CFD simulation and laboratory tests are used to analyze and compare the influence of each solution in detail, and it is successfully applied to the aircraft and verified.

2. Study on solutions

2.1. CFD simulation

The Upper and lower flapper mixer, twin-cyclone mixer and louver mixer are shown in Figure 3.

CFD simulation was carried out on the three mixers to analyze their influence on the uniformity of PCE outlet temperature, the temperature drop (ΔT) between the inlet of the air conditioning system and the wing anti-ice system, and the pressure drop from the PCE outlet to the inlet of air conditioning system(ΔP) with one/two engine bleed systems on while wing anti-ice system is operative. The bleed flow and the PCE cold side temperature were achieved from flight test data. The calculation model based on the system installation of certain large passenger aircraft is shown in Figure 4. The calculation cases are shown in Table 1, and the calculation results are shown in Table 2.

(a) Upper and lower flapper mixer (b) Twin-cyclone mixer (c) Louver mixer

Figure.3 Three kinds of mixers

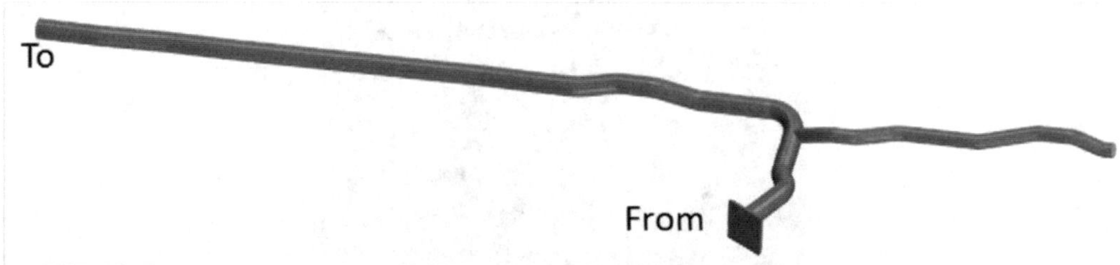

Figure.4 CFD calculation model

Table.1 CFD calculation cases

Cases	Configurations	Bleed airflow	PCE hot side inlet temperature	PCE cold side flow	PCE cold side temperature
		kg/s	°C	kg/s	°C
1	2B2P+WAI	0.938	425.5	0.588	-12.5
2	LBLP+WAI	1.524	416.6	1.138	-0.9

Table.2 CFD calculation result

Mixer	Case	ΔT(°C)	Temperature difference on PCE outlet (°C)	ΔP(psi)
Upper and lower flapper mixer	2B2P+WAI	-1.8	39.3	0.65
	LBLP+WAI	3.7	30.7	1.97

Twin-cyclone mixer	2B2P+WAI	-1.9	29.4	0.52
	LBLP+WAI	2.9	28.9	1.52
Louver mixer	2B2P+WAI	-1.2	37.3	0.54
	LBLP+WAI	3.5	30.9	1.64

As can be seen from Table 2, after adding an upper and lower flapper mixer at the outlet of the PCE, the maximum temperature difference between the inlet of the air conditioning system and the anti-ice system is about 4°C, and the maximum pressure drop between the outlet of the PCE and the inlet of the air conditioning pack is about 2 psi; for the twin-cyclone mixer, the maximum temperature difference between the inlet of the air conditioning system and the anti-ice system is about 3°C, and the maximum pressure drop between the outlet of the PCE and the inlet of the air conditioning pack is about 1.5psi; for louver mixer, the maximum temperature difference between the inlet of the air conditioning system and the anti-ice system is about 3.5°C, and the maximum pressure drop between the outlet of the PCE and the inlet of the air conditioning pack is about 1.6psi. Compared with the other two mixing devices, the overall effect of the upper and lower flapper mixing device is slightly worse.

2.2. Laboratory test

Laboratory tests were carried out to verify the solution of three different mixers mentioned above. As the mixing effect and pressure drop of the upper and lower flapper mixer were slightly worse than those of the other two, only the twin-cyclone mixer and louver mixer were tested in the laboratory.

During the laboratory test, a mixer was installed at the outlet of the PCE, five test conditions under different bleed configurations with different bleed air temperature and airflow were carried out, and several temperature sensors and pressure sensors were set to measure the bleed air temperature and pressure at a different location. The test bench was shown in Figure 5. Temperature differences between the inlet of the air conditioning system and the wing anti-ice system (ΔT) and pressure drop between the outlet of the PCE and the inlet of the air conditioning pack (ΔP) were calculated after the test. The test cases and test results are shown in Table 3.

Figure.5 Test bench

Table.3 Test cases and results

Case	Configurations	Pack flow	Anti-icing flow	PCE hot side inlet temperature	ΔT	ΔP
		kg/s	kg/s	°C	°C	psi
1	2B2P+WAI	0.545	0.345	432	-	-

Case	Configurations	Pack flow	Anti-icing flow	PCE hot side inlet temperature	ΔT	ΔP
		kg/s	kg/s	°C	°C	psi
	Twin-cyclone mixer	0.527	0.356	432	-1.2	0.455
	Louver mixer	0.545	0.343	430	2.2	0.46
2	2B2P+WAI	0.563	0.345	418	-	-
	Twin-cyclone mixer	0.58	0.336	418	-1.2	0.50
	Louver mixer	0.566	0.341	420	2.6	0.47
3	2B2P+WAI	0.575	0.521	402	-	-
	Twin-cyclone mixer	0.549	0.499	405	-0.5	0.45
	Louver mixer	0.627	0.497	400	1.5	0.49
4	1B1P+WAI	0.748	0.345	439	-	-
	Twin-cyclone mixer	0.785	0.345	442	-2	2.26
	Louver mixer	0.736	0.345	438	0.7	1.6
5	1B1P+WAI	0.748	0.345	416	-	-
	Twin-cyclone mixer	0.797	0.344	416	-2	2.26
	Louver mixer	0.739	0.334	420	0.7	1.7

As can be seen from the data in Table 3, the mixing results of both mixers are excellent, the temperature difference is within ±2°Cand the pressure drop is within 2.5psi, which are also in good compliance with the CFD simulation results. These two-mixer designs can effectively solve the problem of temperature stratification at the outlet of PCE and reduce the temperature difference between the air conditioning system and the anti-ice system.

3. Flight Test

Considering the manufacturing process and manufacturing cost, a louver mixer was applied on a certain large passenger aircraft at the outlet of PCE and performed flight tests. During the flight test, the temperature difference between the air conditioning system and the anti-ice system (ΔT) was analyzed. The test results are shown in Table 4.

Table.4 Test data at steady cruising phase

Case	Altitude/ft	Total temperature/°C	ΔT/°C	Configurations
1	5000	14.5	4	2B2P+WAI
2	10000	6.9	4.6	2B2P+WAI
3	16000	-7.3	4	2B2P+WAI
4	18000	-8.3	5.1	2B2P+WAI
5	22000	-15.8	6.1	2B2P+WAI
6	26000	-23.2	6	2B2P+WAI
7	31000	-25.2	6.6	2B2P+WAI

Case	Altitude/ft	Total temperature/°C	ΔT/°C	Configurations
8	35000	-34.4	6.1	2B2P+WAI
9	5000	15.8	3.7	1B1P+WAI
10	10000	12.5	4.7	1B1P+WAI
11	16000	-7.6	4.7	1B1P+WAI
12	18000	-3.6	6.1	1B1P+WAI
13	18000	-4.8	4.9	1B1P+WAI
14	22000	-14.4	4.9	1B1P+WAI
15	26000	-19.8	5.3	1B1P+WAI
16	31000	-25	6	1B1P+WAI

It can be seen from Table 4 that the inlet temperature difference between the air-conditioning system and the wing anti-icing system is ±5°C. The mixer can obviously improve the uneven temperature distribution and reduce the temperature difference between the air conditioning system and anti-ice system to ensure that the temperature supplied to the wing anti-ice system is within the design requirement.

4. Conclusion

This paper creatively proposes design ideas of the three kinds of mixers and determined the solution of adding a mixer downstream of the PCE to solve the temperature deviation problem. The mixer is installed at the hot side outlet of the PCE and used to mix the temperature field at the PCE outlet so that the air supplied to the downstream customer systems meets the design temperature requirements. According to the simulation calculation and laboratory test verification results, the design scheme of a certain large passenger aircraft is selected, which has successfully solved the problem of temperature stratification at the outlet of the PCE, and supported the follow-up flight test.

References

[1] Engine Bleed Air System for Aircraft. SAE ARP 1796B,2015.
[2] Zhang Xu. Failure and maintenance analysis of aircraft engine bleed air system [J]. Internal comb ustion engine and Accessories, 202 (23):161-162.
[3] Jiang Lei. Failure analysis of low pressure of left engine bleed air of aircraft[J]. Jiangsu airlines, 2019 (4):43-44.
[4] Yu Zujian Principle and design of heat exchanger.[M]. Version 2. Beijing: Beijing University of Aeronautics and Astronautics Press, 2019:1-2.

Research on Beam Waveguide Focus Phase Calibration Method of Deep-space TT&C System

Xiaofei Xu[1, 2], Hongjun Hu[2], Shuhong Fu[2], Yalong Yan[2]

[1]State Key Laboratory of Astronautic Dynamics, Xi'an, China.

[2]Xi'an Satellite Control Center, Xi'an, China.

5801267@qq.com

Abstract: The angle auto-tracking capability is needed in the deep space system, but in the sum-differential dual channels tracking model, the deep-space TT&C system must calibrate its phase before tracking a satellite. Because of the big aperture and high frequency of the deep-space TT&C system antenna, it is impossible to build a calibrating tower which satisfy the far-field condition, and the traditional phase calibration method relies on the tower is not suitable. The radio star calibration method selects a appropriate satellite as the calibrating signal source, but the signal of which is very weak and is affected easily. In this paper, the principle of phase calibration is analyzed, and the beam waveguide focus calibration method is studied, the processes of antenna tracking and angle error signal demodulation are given. The beam waveguide focus calibration method simulates the far field signal through controlling the position of horn antenna which is at the focus of the beam waveguide. The method can be realized easily in engineering, and the limit of the traditional calibration tower and the affections of the weather in using radio star calibration method are avoided.

1. Introduction

Aerospace TT&C usually use a single-pulse angle tracking system, which requires phase correction of the equipment before the TT&C tasks. The accuracy of the phase correction directly affects the angle measurement accuracy of the TT&C system. At present, the domestic TT&C system mainly adopts the traditional calibration tower for phase calibration, that is, the calibration tower is built near the TT&C station, and the corresponding calibration equipment is installed on the tower to complete the daily calibration work. Phase calibration with towers has the advantages of good real-time performance and high stability of phasing results, but the construction of calibration towers needs to meet the far-field conditions of the antenna ($R = 2D^2/\lambda$); at the same time, in order to reduce multipath interference, the elevation angle of the antenna to the tower is generally 3 times the antenna beam width, which puts forward certain requirements for the tower height [1]. The deep space TT&C system has the characteristics of large antenna diameter and high working frequency band. The following takes 35m and 64m diameter antennas as examples to calculate the distance and height of the calibration tower to meet the requirements of far-field conditions (calibration elevation angle 3°), as shown in Table 1.

Table 1 Far-field conditions of deep space TT&C system

Antenna diameter (m)	Frequency (MHz)	Far field condition (m)	Tower height (m)
35	2300	18783	1008
35	8500	69417	3658
35	32300	263783	13830

64	2300	62805	3327
64	8500	232107	12188
64	32300	882005	46201

It can be seen from Table 1 that it is not feasible for the deep space TT&C system to build a calibration tower under the far-field condition to complete the phase calibration with the tower. Therefore, in order to solve the phasing problem of the deep space TT&C system, it is necessary to carry out the research on the towerless phasing method [2]. The existing geostationary satellite resources are also difficult to meet the pre-mission angle calibration needs in practical applications. Therefore, the angle tracking system and the phase consistency calibration of the difference channel have become a technical problem of the deep space TT&C system. References [3-5] analyzed the radio satellite phasing method in detail, and gave the experimental verification results. Reference [6] proposes a real-time correction method for fast phasing of stars. References [7, 8] studied the method of using UAV platform for phasing. The focus phasing method studied in this paper is to place a beacon horn at the focal position of the beam guide, so that the focus of the beacon horn coincides with the focus of the beam guide, which is equivalent to the signal generated in the beam guide under far-field conditions. By offsetting the position of the horn, phasing can be performed by simulating the conventional phasing process.

2. Basic principles of phasing

When the antenna receives circularly polarized waves, the tracking coupler outputs a set of orthogonal TE_{11} modulated signals, which are the difference branch azimuth signal ΔAZ and the difference branch elevation signal ΔEL respectively, with a phase difference of 90°.

In a dual-channel tracking receiver, the general definition and signal are

$$u_{\Sigma(t)} = \cos(\omega t + \phi) \quad (1)$$

The difference branch signal (left-handed) is

$$u_{\Delta(t)} = A\cos(\omega t + \phi_1) + E\sin(\omega t + \phi_2) \quad (2)$$

The difference branch signal (right-handed) is

$$u_{\Delta(t)} = A\cos(\omega t + \phi_1) - E\sin(\omega t + \phi_2) \quad (3)$$

where ϕ is the phase of the sum signal, ϕ_1 and ϕ_2 are the phases of the difference signal, A is the amplitude of the in-phase component of the difference signal, and E is the amplitude of the quadrature component of the difference signal. Taking the left-handed signal as an example, after the tracking receiver and the path are locked, the local tracking receiver NCO passes through the phase shifter to generate the reference signal of the differential path detector: azimuth $k_1\cos(\omega t + \phi + \theta_1)$, pitch $k_2\sin(\omega t + \phi + \theta_2)$. Among them, k_1 and k_2 are the gain coefficients of the azimuth and pitch branches, and θ_1 and θ_2 are the phase shift values of the azimuth and pitch branches. The phasing process is to obtain θ_1 and θ_2 so that they satisfy:

$$\phi + \theta_1 = \phi_1, \phi + \theta_2 = \phi_2 \quad (4)$$

When the antenna difference branch pattern satisfies the ideal center symmetry, $\theta_1 = \theta_2$. The strength of the difference branch signal determines the angle that the target deviates from the direction of the electrical axis of the antenna. In order to eliminate the signal energy change caused by the distance of the target, it is necessary to normalize the energy of the signal of the difference branch and the signal of the channel.

3. Beam waveguide focus phase correction method

3.1. Angle tracking process

When the Cassegrain beam waveguide antenna monopulse self-tracking system works, the spatial angle where the incoming wave direction deviates from the main axis of the antenna, and is projected onto the pitch axis and the cross pitch axis of the antenna. After correct phase correction, the demodulation output error voltage reflects the projection state of the incoming wave direction on the pitch axis and the cross pitch axis.

When the direction of the incoming wave is parallel to the direction of the electrical axis, the projection on the azimuth axis and the pitch axis is zero, the output of the feed difference signal is zero, and the electrical axis of the sum signal is aligned with the target and the output is the largest. When the incoming wave direction and the antenna electrical axis have a certain deviation angle, the signal has a projection component $V_d\cos(\omega t)$ on the azimuth axis and the elevation axis plane of the antenna aperture. After the projection component is decomposed by the vector, the azimuth axis error signal projection $V_a\cos(\omega t)$ and the elevation axis error projection $V_e\cos(\omega t)$ are formed on the azimuth axis and the elevation axis of the antenna surface. For the antenna to perform self-tracking, the tracking receiver needs to demodulate V_a and V_e, and the antenna servo system adjusts the direction according to the error voltage to complete the target tracking.

If the far-field static beacon conditions are available, the conventional method can be used for phase calibration, that is, after the antenna electrical axis is aligned with the beacon, only the azimuth (or pitch) is offset by a known angle to complete the phase calibration and directional sensitivity calibration. The general process of conventional phasing is as follows: the antenna electric axis is aligned with the calibration tower electric mark, the azimuth and pitch are deviated by a certain angle once, and the phase value and channel gain of the tracking receiver phase shifter are manually adjusted to make the output error voltage amplitude and polarity meet the requirements; after the four-quadrant bias adjustment, if the polarity, sensitivity and cross-coupling meet certain conditions, the antenna is closed-loop and self-tracking, and a new tracking zero point is obtained; then re-check the polarity, sensitivity and cross-coupling with the new zero point. If the requirements are met, the phasing ends; if not, the phasing is performed again until the requirements are met.

3.2. Angular error signal demodulation

The error signal Vd*cos(wt) is irradiated on the receiving feed after passing through the main surface, the auxiliary surface and the beam waveguide. The error signal excites a higher-order mode (TE21 mode) in the waveguide system in the receiving feed. There are higher-order mode couplers in the receive feed. A coupler is formed by four coupling arms and a summer. The signals output by the two couplers are combined with a 90-degree quadrature sum circuit to synthesize the difference signal. The two couplers establish an orthogonal coordinate system in the circular waveguide, which decomposes the higher-order mode error signal Vd*cos(wt). After decomposition, the error signals Vx*cos(wt) and Vy*cos(wt) are obtained. After 90 degrees and the circuit, the signal becomes:

$$\begin{aligned} V_{\Delta}(t) &= V_x\cos(\omega t) + V_y\cos(\omega t) \\ &= \sqrt{V_{X_x}^2 + V_{Y_x}^2}\left(\frac{V_x}{\sqrt{V_{X_x}^2 + V_{Y_x}^2}}\cos(\omega t) + \frac{V_y}{\sqrt{V_{X_x}^2 + V_{Y_x}^2}}\cos(\omega t)\right) \\ &= \sqrt{V_{X_x}^2 + V_{Y_x}^2}\cos(\omega t + \theta) = V_d\cos(\omega t + \theta) \end{aligned} \tag{5}$$

where $\theta = arctg(\frac{V_x}{V_y})$ represents the deviation angle of the synthesized error signal relative to the orthogonal coordinate system of the higher-order mode coupler.

The signal is decomposed in quadrature by the coherent demodulation unit of quadrature local oscillator in the tracking receiver, and decomposed into two signals Va (t) and Ve (t). The orthogonal decomposition process of the demodulation unit of the tracking receiver obtains Va (t) (Ve (t)), and the antenna aperture is decomposed according to the antenna azimuth axis and the elevation axis to obtain Va*cos(wt) (Ve*cos(wt)), which not necessarily parallel, that is, Va(t) is not only related to Va*cos(wt), but also related to Ve*cos(wt), expressed as:

$$Va\ (t)=f(Va*\cos(wt),\ Ve*\cos(wt)) \tag{6}$$

3.3. Focus correction process

There is a focal point in the deep space antenna beam waveguide, and the far-field problem can be solved by using the focal point. As shown in Figure 1, a beacon horn is placed at the focal position, so that the focal point of the beacon horn coincides with the focal point of the beam guide, that is, the signal generated in the beam guide under the far-field condition can be equivalent.

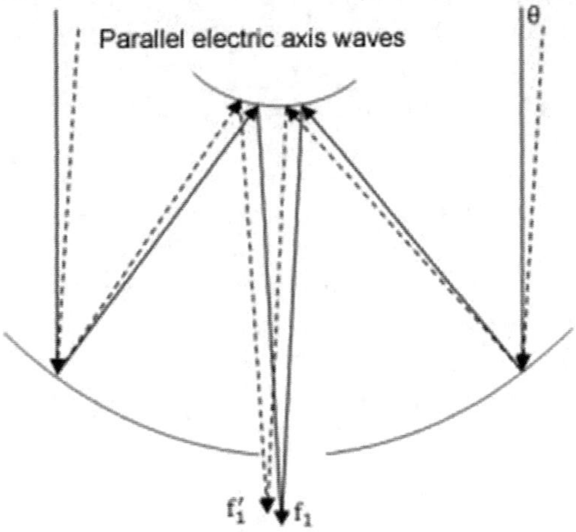

Figure 1 Simulated far-field signal with change of beam guide focal point

When using the conventional phasing method, the antenna is controlled to be offset only in the azimuth axis (or only in the elevation axis), so that the electric axis of the antenna is deviated from the target by a known angle. In this case, the focal position of f_1 changes accordingly, the focal position becomes f_1'. Or in the differential phase method, f_1 is the focus position at the first angle, and f_1' is the focus position at the second angle. If the position of the beacon horn can be controlled so that its focus can be moved between the focus f_1 and the focus f_1', the situation of the far-field beacon performing conventional phasing can be completely simulated. Under far-field conditions, the positional relationship between the angle θ of the target deviating from the electrical axis in the azimuth axis (or in the elevation axis) and the positional relationship between the focus f_1 and the focus f_1' are determined by the geometry of the main and sub-reflecting surfaces of the antenna, and can be obtained by geometric calculation. Controlling the movement of the horn position at the f_1 focus can be accomplished by a three-axis precision mount for x, y, and z. The moving accuracy of the horn reaches 1/100 of the wavelength of the received signal, which can accurately find the position of f_1 and complete the phase correction process of the conventional method.

4. Conclusion

In this paper, the principle of the conventional phasing method is analyzed, the beam-waveguide focus phasing method is studied, and the tracking and angular error signal demodulation process of the wave-speed waveguide antenna is given. The beam reporting focus phasing method is a new beam waveguide antenna phasing method, which solves the problem that the deep space TT&C system cannot use the traditional calibration tower phasing due to the large antenna diameter and high frequency band, and the design idea is ingenious, simple and feasible .

In the next step, we will carry out the experimental verification of the beam-waveguide focus phasing method in combination with the project.

References

[1] Binfeng Yang. Angle Calibration Technique of Ground -based Telemetry and Control Radar [J]. Modern Electgronics Technique, 2007, (17): 47-52.
[2] Jianzhong Shang, Qingsong Wang, Xinhong Wang. A Towerless Phase Calibration Method Based on Record of Error Voltage Ratio [J]. Telecommunication Engineering, 2015, (9): 1014-1018.
[3] Chan Li, Min Liu, Yinong Yu. Research on Phase Calibration Using Radio Star for Deep Space TT & C System [J]. Journal of Spacecraft TT & C Technology, 2010, 12(6): 28-30.
[4] Sanshan Qiu, Yuanling Wang, Hongjun Yang. Feasibility Analysis of Phase Calibration for Tracking Receiver Using Radio Star in Deep Space TT&C System [J]. Telecommunication Engineering, 2010, 50(8): 22-25.
[5] Jiaxing Liu. Towerless Phase Calibration Using Radio Star Noise [J]. Telecommunication Engineering, 2010, 50(6): 1-4.
[6] Jin Hou. A method of quickly adjust phase between sum and diff channel in large aperture parabolic antenna by satellite based on real-time guide angle correction [J]. Journal of Telemetry, Tracking and Command, 2022, 43(2): 91-96.
[7] Yu Hong, Zongqing Wu, Tao Men, et al. Design and Test Verification of Phase Calibration Mode of Ka-Band Large Aperture TT&C Antenna Based on Unmanned Aerial Vehicle [J]. Journal of Astronautics, 2021, 42(6): 783-789.
[8] Ming Lei. Design and Application of TT&C Ground System Phase Calibration Based on UAV Platform [J]. Electronics Quality, 2021, (4): 50-53.

GIS-based Ecological Risk Assessment on Land Use In Chongqing City, China *

Ziyue Wang†

School of Earth and Environmental Science, the University of Queensland, Brisbane QLD, 4072, Australia

carina.ziyue@gmail.com

Abstract—Ecological risk assessment (ERA) is a process of predicting the possibility of harmful effects of environmental pollutants on an entire or part of the ecosystem, which is closely connected with the early human health risk assessment. A rational assessment of ecological risks in an urban area can provide great significance for urban landscape pattern optimization, urban ecological risk warning mechanism establishment, and urban ecological environment risk reduction. This study is based on GIS system to assess the ecological risk of land use in Chongqing in 2020, according to the basic theoretical framework and method system of ecological risk assessment. The main process is using remote sensing image data to construct the Ecological Risk Index (ERI) from the perspective of landscape patterns to reflect the potential impact of resource-based urban land use on the ecological environment. In addition, the ArcGIS spatial analysis method was used to evaluate Chongqing's ecological risk. As a result, high-risk area close to the water body just occupied a small part of Chongqing City (around 5%) and this area should be taken into protective measures urgently.

1. Introduction

Ecological risk refers to the normal function of a population, ecosystem or the entire landscape that is subject to external stress (Makokha et al.,2016)[1]. ERA is a process of predicting the possibility of harmful effects of environmental pollutants on an entire or part of the ecosystem, which is closely connected with the early human health risk assessment. The development of the ERA process is gradually developed from environmental impact assessment and environmental risk assessment (Zhao et al.,2014)[2]. ERA reflects the negative impact of one or more factors on the ecosystem and can characterize the threats they pose to the ecosystem from climate change and human behaviors (Lu et al.,2017)[3]. ERA was first formally defined in the 1980s and the research mainly concerned with ecological pollution. In recent years, the spatial distribution of ecological risks and their impact on ecosystem services have been two popular research topics in ecology (Walker et al.,2001)[4]. The scale of assessment has expanded from a single organism to a watershed, regions and cities. Regional ecological risk assessment evaluated the threat degree of various adverse effects on ecosystems exposed to possible ecological hazards in the ecological system.

At present, regional ecological risk research is mainly concentrated in arid ecologically fragile areas in the west of China, but fewer types of research are based on an urban area. Chongqing is a typical mountainous city in southwestern China and a special area under many important policies. For instance, the Three Gorges Dam was established in 1993 and operated in 2005 and the Western China Development Program has made a huge impact on Chongqing's land use changes over the past 30 years. Although these policies are beneficial to fueling the socioeconomic development of Chongqing, the ecological problems (e.g., desertification and river basin pollution) have become increasingly serious with the rapid development of the city.

Based on the above background, This study is based on GIS system to assess the ecological risk of land use in Chongqing in 2020, according to the basic theoretical framework and method system of ecological risk assessment. The main process is using remote sensing image data to construct the ERI from the perspective of landscape patterns to reflect the potential impact of resource-based urban land use on the ecological environment. In addition, the ArcGIS spatial analysis method was used to evaluate spatial characteristics. This project can provide a quantitative decision basis for strengthening the ecosystem in Chongqing

2. Method

2.1. Study site

In this project, the study area is Chongqing which is located in southwestern China, The longitude of Chongqing area is 105°11'-110°11' E and the latitude is 28°10'-32°13' N. The main cover land is a mountainous area, hill, and Yangtze River valley (Figure1). The climate in the study area is a humid subtropical monsoon climate, which is beneficial for agriculture.

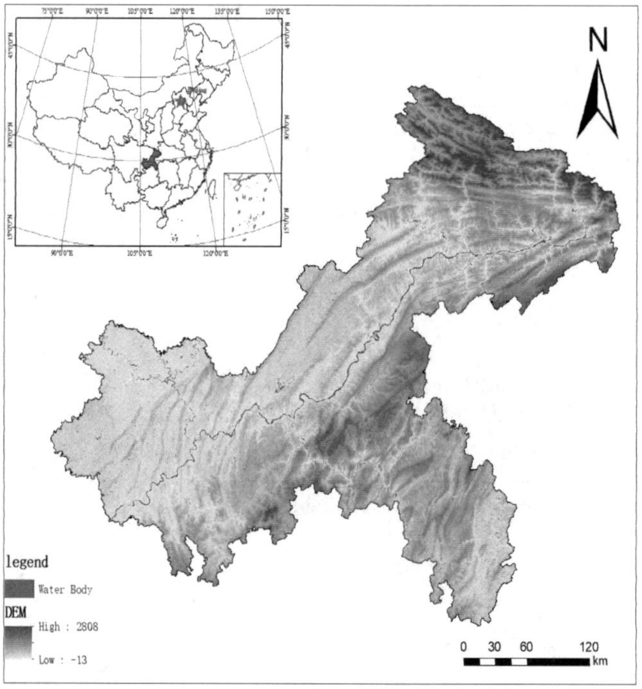

Figure 1: The geographical location of Chongqing

2.2. Data site

The main data set includes shapefile data of the boundary of Chongqing City, DEM, and related data on land use. The raw material of land use data is the world's 30-meter global surface coverage product V1.0 with National Standard for Land Use Status Classification (GB/T21010—2007). It can be downloaded from the open-source platform of the Institute of Aerospace Information Innovation, Affiliated with Chinese Academy of Sciences. Based on the boundary data of Chongqing, it can be found that the raw land use data can be cropped and the final format land use data shows 6 landscape types. (As shown in Figure 2).

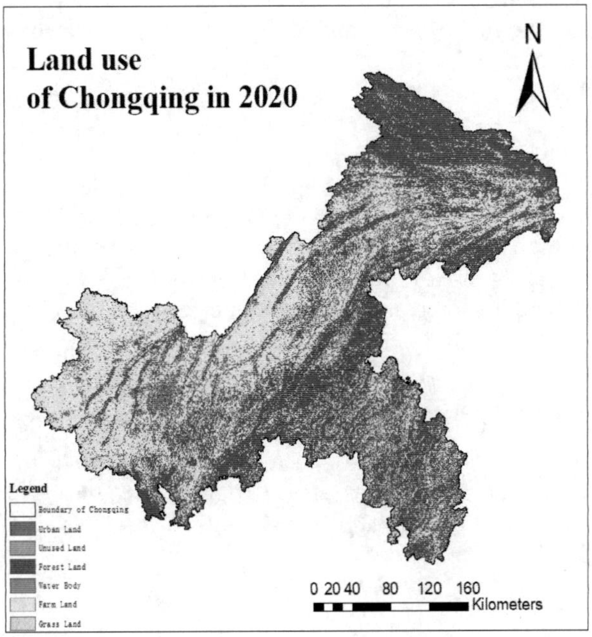

Figure 2: Land use of Chongqing in 2020

2.3. Landscape ecological risk index

2.3.1 Establish risk evaluation cell. To spatialize the ecological risk index, a risk evaluation cell should be created first. Based on the scope and considering the workload of the study area, there are 979 cells (10×10km) created in the study area through the Data Management tool in ArcGIS Pro (Figure 3). The value of each cell is calculated by the ecological risk index.

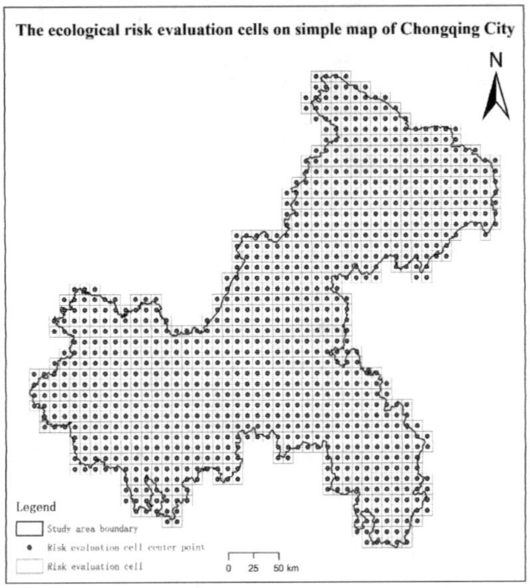

Figure 3: Ecological risk evaluation cells on a sample map of Chongqing City

2.3.2 Calculate landscape ecological risk index. To establish the link between regional ecological risk and land use types in 2020, the ERI of each land use type is calculated. The ERI calculation formulas are established below:

$$ERI_i = \sum_i^N \frac{A_{ki}}{A_k} R_i \quad (1)$$

$$R_i = U_i * ES_i \quad (2)$$

Where ES_i is the Landscape fragility. Combined with the expert's method of estimating scores, the fracture degree values of the study site are divided into 6 levels from high to low: Unused land is scored as 6, Water land is scored as 5, Cultivated land is scored as 4, Grassland is scored as 3, Forest land is scored as 2, Urban land is scored as 1. U_i is the interference index of the i-th landscape and the formula U_i is below:

$$U_i = aC_i + bF_i + cD_i \quad (3)$$

Where C_i is landscape fragmentation. F_i is landscape isolation, D_i is landscape dominance index. a, b, and c is the different degrees of influence of each index. According to previous research (Li et al.,2017), C_i is the most important and a is assigned to 0.6, followed by F_i and D_i and values are 0.3 and 0.1 respectively. The formula of C_i, F_i and D_i are below:

$$C_i = \frac{N_i}{A_i} \quad (4)$$

$$F_i = \frac{\sqrt{S_i}}{2P_i} \quad (5)$$

$$D_i = dL_i + eP_i, \quad (6)$$

$$L_i = \frac{N_i}{N} \quad (7)$$

Where N_i is the number of landscape type patches, N is the total number of landscape cells, L_i is the relative density of the landscape type, and P_i is the relative coverage of the landscape type.

2.3.3 Spatial analysis of ecological risk. To realize regional ecological risk analysis, the center point of 200 cells is assigned to the ecological risk index value through the Geostatistical formula:

$$\gamma(h) = \frac{1}{2N(h)} \sum_i^{N(h)} [Z(x_i) - Z(x_i + h)]^2, (i = 1,2,3, ..., N(h)) \quad (8)$$

In the Geostatistical formula, $\gamma(h)$ is the variation function, h is the step length which is the spatial separation distance of paired sampling, $N(h)$ is the number of sample points when the separation distance is h.

Then, based on the geostatistical analysis model (Histogram model) in the ArcGIS pro software, the ERI value is close to a normal distribution and the data has a spatial correction. Therefore, the ordinary Kriging method is used to interpolate the ecological risks of Chongqing City. Finally, the ecological risk map of the study area would be distributed.

3. Results

3.1. Analysis of land use structure characteristics

As Figure 2 shows, there are 6 landscape classifications. Based on a vector to raster tool and calculate geometry method in ArcGIS Pro, the whole area of Chongqing City is $8245.01km^2$. Forest land dominates half the land of Chongqing around $42698.26km^2$ and farmland occupies almost half the land of Chongqing (42%). However, the occupation of grassland, urban land, water body and the unused area is less, respectively around 1.9%, 2.6%, 1.1% and 0.4%. The details of each land use occupation are collated in the following table (As shown in Table 1).

Table 1: Area of each landscape and Area proportion of ecological risk for different patterns in 2020

Land use type	Farm Land	Forest Land	Grass Land	Urban Land	Water Body	Unused Land
Area/km^2	35108.38	42698.26	1601.50	2138.21	938.48	0.18
Area Proportion/%	42	52	1.9	2.6	1.1	0.4

3.2. Calculation results of parameters in the ecological risk index

The separation degree (F_i) of each landscape type in 2020 was: Unused land (728.27) > Grassland (4.90) >Urban land (2.77) > Water body(1.79)>Farm land (0.23)>Forest land (0.19). This means the landscape separation degree of various types of landscapes is very different. Among them, the Unused landscape has the largest separation degree and indicates that the Unused landscape has the widest and the most scattered distribution in the landscape, while the Farm landscape has the smallest separation degree. Similarly, the Landscape interference index (U_i) of Unused landscape is the highest (221.25), which means the unused landscape is easier to be affected by natural and human factors.(As shown in Table 2)

Table 2: Parameters related to the ecological risk index

Land use type	N_i	A_i/km^2	C_i	S_i	P_i	F_i	L_i	D_i	U_i
Forest land	319515	4260760.47	0.70	0.03	0.52	0.19	0.27	0.37	0.17
Farmland	307509	3498537.51	0.09	0.03	0.42	0.23	0.26	0.33	0.18
Urban land	184203	222197.40	0.83	0.02	0.03	2.77	0.16	0.11	1.27
Grassland	347427	173009.88	2.00	0.04	0.02	4.90	0.30	0.19	2.51
Waterbody	13686	94083.93	0.15	0.001	0.01	1.79	0.01	0.01	0.61
Unused land	119	21.51	5.53	0.00001	0.000003	728.27	0.0001	0.00006	221.25

3.3. Visualization of ERI value

According to Figure 4, the degree of ecological risk in Chongqing City can be clearly divided into three levels, namely high, medium and low ecological risk areas. Areas far away from the central city, especially in the western part of Chongqing City, are relatively forest and grassland, with a small proportion of industrial, mining, residential and transportation land, and have little impact on the ecological environment. Its ecological risk index value is low, the highest risk area is located east of Chongqing.

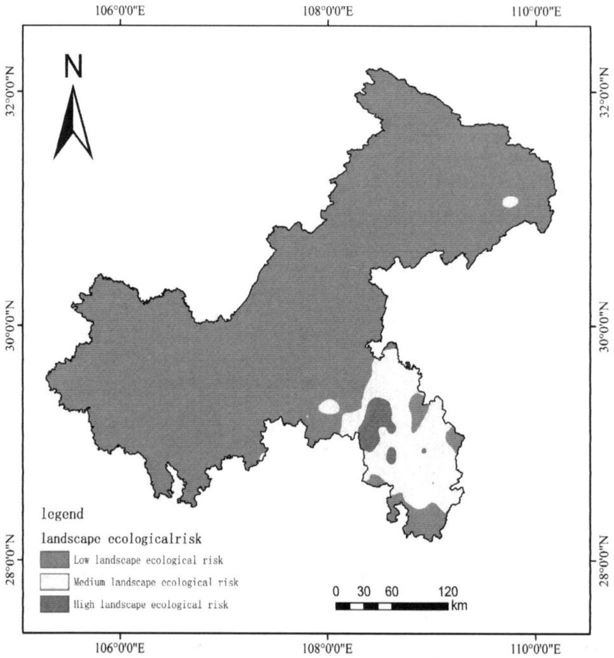

Figure 4. Landscape ecological risk map of Chongqing in 2020

4. Discussion

The ecological risk index is a comprehensive probability measurement relative to the possible occurrence of ecological problems and it cannot provide a direct basis for specific ecological protection decisions. However, based on the degree of ecological risk, The ecologically high-risk areas derived from this study are undoubtedly the areas that need to be paid more attention to in the future ecological protection work by improving land production functions, especially near a water body, and strengthening the construction of urban land. Ecological construction in medium and low ecological risk areas is also an issue that cannot be ignored. It is the goal of ecological construction to strive to build the land into a land production and living environment with reasonable structure, sound functions, a beautiful environment, and increasing economic benefits.

5. Conclusion

This study with the strong support of GIS technology, starts from the factors affecting ecological risk to determine which areas are prone to disasters and the probability of occurrence. Based on the land use characteristics of Chongqing area, the analysis constructs an ecological evaluation index system in line with the area and evaluates the ecological risk level of Chongqing area. The constructed ecological risk evaluation system overcomes the shortcomings of traditional single-factor evaluation, comprehensive index evaluation and ecological risk index evaluation, and provides a new research method for ecological risk assessment. The low-risk area is mainly the forest and grass area. The medium-risk area is located in the northeastern area where mainly farmland. The high-risk area is an urban area and is close to a water body. Fortunately, the proportion of high-risk areas just occupies a small part of the whole city, around 5%. This influence that the policy of protecting the environment is efficient.

Acknowledgments

All the data are downloaded from open-source websites.

References

[1]Makokha, V. A., Qi, Y., Shen, Y., & Wang, J. (2016). Concentrations, Distribution, and Ecological Risk Assessment of Heavy Metals in the East Dongting and Honghu Lake, China. Exposure and Health, 8(1), 31-41.

[2]Zhao, Y., Huang, C.-M., & Zhang, H.-J. (2014). Ecological Risk Assessment of Provincial Land-Use Overall Planning in China. Human and Ecological Risk Assessment, 20(6), 1491-1506.

[3]Yan Lu, Fen Qin, Zhongbing Chang, & Shuming Bao. (2017). Regional Ecological Risk Assessment in the Huai River Watershed during 2010-2015. Sustainability (Basel, Switzerland), 9(12), 2231.

[4]Walker, R., Landis, W., & Brown, P. (2001). Developing A Regional Ecological Risk Assessment: A Case Study of a Tasmanian Agricultural Catchment. Human and Ecological Risk Assessment, 7(2), 417-439.

Identification of Burned Areas Based on Texture Features of Different Windows

Jinwen Wu[1], Yushu Zhang[1], Longyu Sun[2], Ruipeng Ji[1], Wenying Yu[1], Rui Feng[1*]

[1]Institute of Atmospheric Environment (IAE), CMA, Shengyang,110166

[2]Meteorological bureau of Shenyang, Shengyang,110168

* Corresponding author: fengrui_k@126.com

Abstract: It is vital to choose appropriate pixel spacing as the window size determines the distribution of texture features in a gray-level co-occurrence matrix (GLCM). The window size ranged from 9×9 to 15×15 to keep the richness of texture image information. In this study, the importance evaluation method based on the Gini index was used to determine the appropriate window for the identification of burned areas, and burned areas were identified with the random forest classification method based on texture features. The results showed that 15×15 was the most appropriate texture window size for the identification of burned areas, and the best texture features for the identification of burned areas were the contrast of the visible blue light band and the texture mean of the near-infrared band; burned areas could be fully identified with the random forest classification method based on texture features, with an overall classification accuracy of 83.87%, but there would be interference pixels in reservoirs, pits, ponds, marshes and irrigated dry fields.

1. Introduction

Textures reflect visual characteristics of homogeneity in images with smoothness, roughness, fineness, granularity, and randomness, which can improve the classification ability of satellites [1-2]. Texture feature extraction methods of remote sensing images include gray-level co-occurrence matrix (GLCM), Markov random field model (MRF), local binary pattern (LBP), Fourier filter (FFT), wavelet transform, etc. [3-4], among which GLCM has been regarded as an effective and common method [5]. GLCM textures are widely used in the identification of forest tree species, health monitoring, stock volume inversion, etc. [6-7] but there are few studies on its application in burned areas. Smith et al. [8] used first-order GLCM textures to extract burned areas based on radar satellites, and the results showed that GLCM textures improved the ability to identify burned areas on low spatial resolution images. Mitri et al. [9] put forward a semi-automated object-oriented model for burned areas based on the Landsat satellite, concluding that textures could separate areas with the same spectral mean and thus avoid the confusion of "different objects with the same spectrum." Liu et al. [3] optimized the texture window conditions of land cover, and the results showed that texture features could reduce the error caused by similar surface reflectivity among different types. All the studies above showed that textures could enhance the detailed features of burned areas and facilitate the distinction of confusing surface features.

2. Research data and method

2.1 Research data

GF-1 and GF-2 remote sensing images on May 11th, 2019 (Figure 1) from China Centre for Resources Satellite Data and Application, as the data sources, were pre-treated with ENVI5.3 in the manners such as radiation correction, atmospheric correction, geometric correction, and image enhancement for the extraction of burned areas. The land use data from the Resource and Environment Science and Data Center, Chinese Academy of Sciences, was used for the forest land extraction.

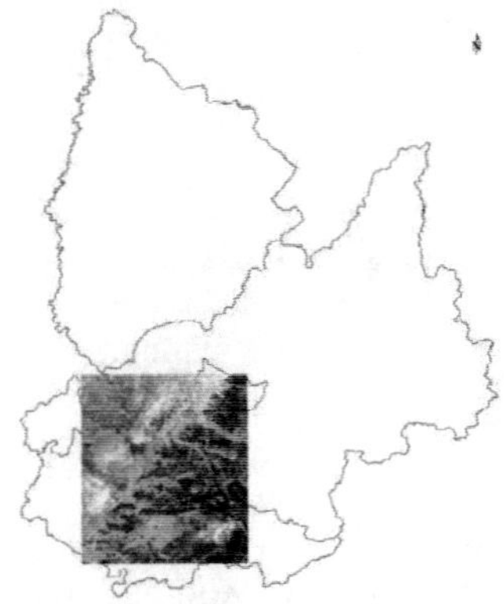

Figure.1 Location of the study area

2.2 Texture calculation

GLCM represents textures by calculating the joint conditional probability density P(i,j) between gray levels of images. P(i,j) indicates the probability of the occurrence of the gray level j with the gray level i as the starting point, which is usually expressed in matrix form, i.e., the gray-level co-occurrence matrix.

$$P(i,j) = \{[(x,y),(x+\Delta x, y+\Delta y)] | f(x,y) = i, f(x+\Delta x, y+\Delta y)\} \quad (1)$$

Where i and j were valued in the range of$\{0,1,2 \ldots\ldots \mathrm{k}-1\}$, and the window sizes were 3×3, 5×5, 7×7, 9×9, 11×11, 13×13, and 15×15. In most studies, the window is set to 3 pixels when the texture value is calculated based on GLCM. How the window size affects the texture value, and the extraction of burned areas has not been widely discussed in recent literature [10].

Table 1 Calculation formula of parameters of GLCM texture features

Serial number	GLCM texture features	Calculation formula
1	Mean	$Mean = \sum_{i,j=1}^{G} i * P(i,j)$
2	Variance	$VAR = \sum_{i=1}^{G}\sum_{j=1}^{G}(i-\mu)^2 P(i,j)$
3	Homogeneity	$HOM = \sum_{i=1}^{G}\sum_{j=1}^{G}\frac{P(i,j)}{1+(i-j)^2}$
4	Contrast	$CON = \sum_{i=1}^{G}\sum_{j=1}^{G}(i-j)^2 P(i,j)$
5	Dissimilarity	$DIS = \sum_{i=1}^{G}\sum_{j=1}^{G} P(i,j) * \lvert i-j \rvert$

6	Entropy	$ENT = -\sum_{i=1}^{G}\sum_{j=1}^{G} P(i,j)\log[P(i,j)]$
7	Angular Second Moment	$ASM = \sum_{i=1}^{G}\sum_{j=1}^{G} P^2(i,j)$
8	Correlation	$COR = \sum_{i=1}^{G}\sum_{j=1}^{G} \frac{(i - Mean_i)(j - Mean_j)P(i,j)}{\sqrt{VAR_i VAR_j}}$

2.3 Importance evaluation method based on Gini index

The texture window sizes are usually 3×3, 5×5, 7×7, 9×9, 11×11, 13×13, and 15×15. Pan et al. also found during the extraction of texture features on high-resolution remote sensing images that the appropriate moving window size ranged from 9×9 to 15×15to keep the richness of texture image information. Therefore, seven moving windows with different sizes (3×3, 5×5, 7×7, 9×9, 11×11, 13×13, and 15×15) were selected in this study, and the appropriate window size for the extraction of burned areas was determined with the importance evaluation method based on the Gini index. The calculation formula is shown below:

$$Gini(P) = \sum_{i=1}^{N} P_i(1 - P_i) = 1 - \sum_{i=1}^{N} P_i^2 \tag{2}$$

Where, i represents *i* categories, and P_i refers to the weight of category *i* in the sample set;

Then, the importance of the feature is $x_j$at node m, i.e., the change of the Gini index at node m before and after the branch is:

$$VIM_{jm}^{Gini} = GI_m - GI_l - GI_r \tag{3}$$

Where, GI_l and GI_r refer to the Gini index at the two new nodes before and after the branch, respectively;

If the node where the feature x_j occurs in the decision tree *i* is in the set m, the importance of x_j in the ith tree is:

$$VIM_{ij}^{Gini} = \sum_{m \epsilon M} VIM_{jm}^{Gini} \tag{4}$$

Assuming that there are n decision trees, the importance of the characteristic variable is:

$$VIM_{j}^{Gini} = \sum_{i=1}^{n} VIM_{ij}^{Gini} \tag{5}$$

3. Conclusion

3.1 Texture features of different moving windows

It is vital to choose appropriate pixel spacing as the window size determining the distribution of texture features in GLCM. Gray-level textures are generated according to the (7) different window sizes. Each texture image containes four channels and eight texture features of GF-2 remote sensing images. A total of 32 texture channels are generated. Figure 2 shows the colored composite diagram of textures with different window sizes (R is the texture mean of channel 1, G is the texture mean of channel 4, and B is the texture contrast of channel 3).

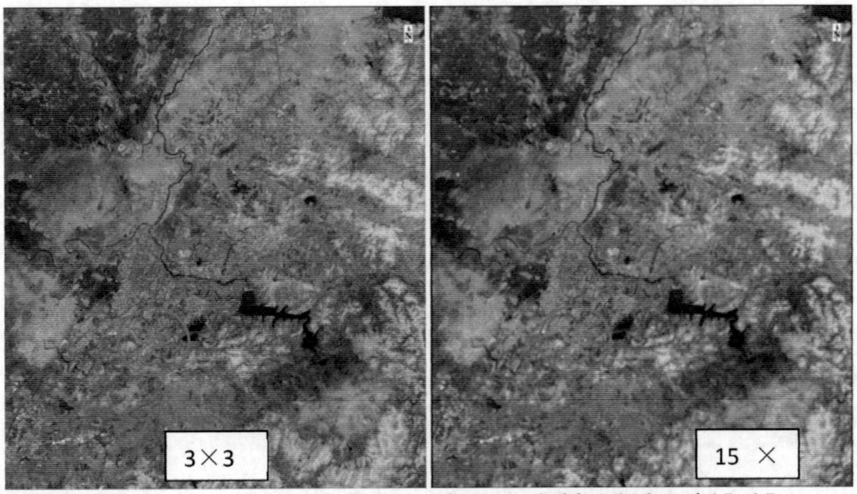

Figure 2 Colored composite diagram of texture features (taking 3×3 and 15×15 as examples)
Note: The red circle mark on the image indicates the burned area.

3.2 Importance of evaluation of texture window parameters

The appropriate window size for the extraction of burned areas was judged according to the Gini index. The results showed that the importance of texture features in the extraction of burned areas increased (red trend line) with the increase of the texture window size and was maximized when the window size reached 15x15, indicating that object-oriented identification was better than pixel-based identification; the most important texture feature was the contrast of the visible light band B1 (blue light 0.45-0.52um), followed by the texture mean of the near-infrared band B4 (0.77-0.89um)and that of the band B1; and the texture means with other window sizes also showed high importance. To sum up, 15×15 was the most appropriate window size for the identification of burned areas, and the best texture features for the identification of burned areas were the contrast of the visible blue light band and the texture mean of the near-infrared band.

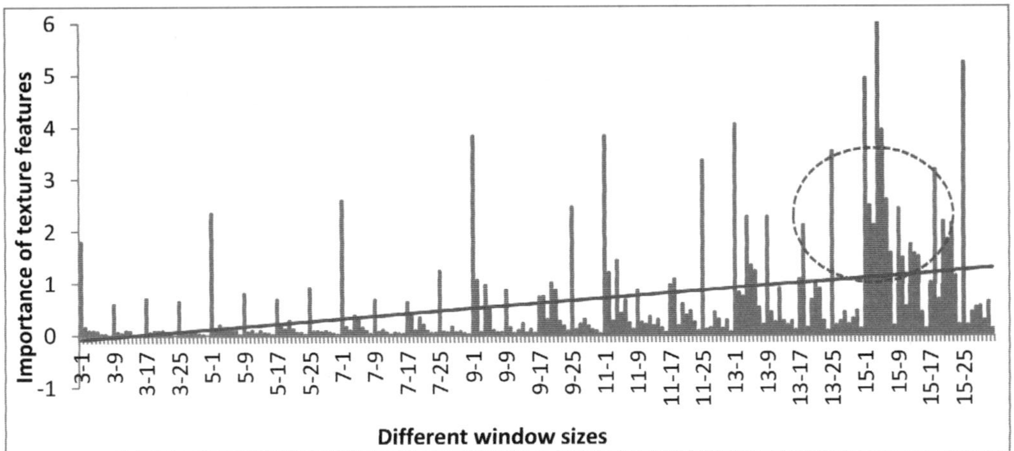

Figure 3 Importance of texture features with different window sizes in classification and extraction of burned areas

3.3 Identification of burned areas based on texture features

The burned areas were identified with the random forest classification method based on texture features (Figure 4). They could be fully identified, but some interference points (dotted line ranges)

were mainly distributed in reservoirs, pits, ponds, marshes, and irrigated dry fields and were greatly affected by water bodies. The burned areas of the forest type were extracted in combination with land use, and the identification accuracy was tested with the confusion matrix method based on the higher-resolution satellite remote sensing image GF-2 (spatial resolution 4m). The overall classification accuracy was 83.87%, and the Kappa coefficient was 0.7451.

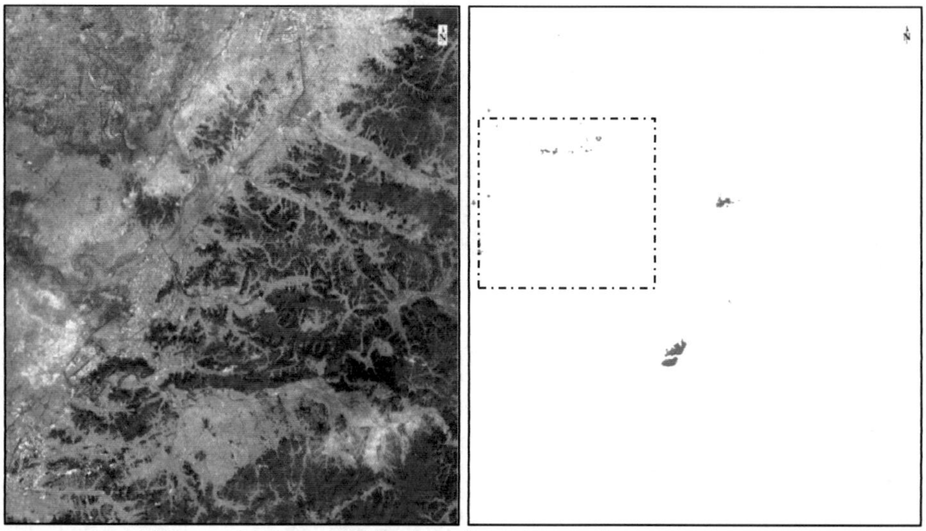

Figure 4 Identification of burned areas

4. Results and discussion

The selection of texture windows is quite important. An excessively large window will lead to different texture elements in the same window, while an excessively small window will cause the lack of a complete texture element in the same window. Therefore, the burned areas were identified only in the range of conventional texture windows (9×9 to 15×15) in this experiment to keep the richness of texture image information. 15×15 was the most appropriate texture window size for the identification of burned areas, indicating that object-oriented identification was better than pixel-based identification; the best texture features for the identification of burned areas were the contrast of the visible blue light band and the texture mean of the near-infrared band; burned areas could be fully identified with the random forest classification method based on texture features, with an overall classification accuracy of 83.87%, but there would be interference pixels in reservoirs, pits, ponds, marshes and irrigated dry fields. To improve the identification accuracy of high-resolution satellite remote sensing images, it is necessary to further study the object-oriented and multi-feature machine learning classification method.

Funding information

This work was financially Co-sponsored by the Shenyang young and middle-aged scientific and technological innovation talents support program (RC210431), FY satellite application advance plan (FY-APP-2021.0302), and the basic scientific research projects of the central public welfare research institutes (2021SYIAEMS1)

References

[1] Mryka Hall-Beyer (2017) Practical guidelines for choosing GLCM textures to use in landscape classification tasks over a range of moderate spatial scales, International Journal of Remote Sensing, 38:1312-1338

[2] Venkateswaran K, Kasthuri N, Alaguraja R A. (2017) A Combined Detail Enhancing Algorithm and Texture Feature Extraction Method for Supervised Classification of Remote Sensing Images[J]. Journal of the Indian Society of Remote Sensing, 46:327–336.
[3] Liu L, Chen J, Fieguth P, et al. (2019) From BoW to CNN: Two Decades of Texture Representation for Texture Classification[J].International Journal of Computer Vision, 127: 74–109.
[4] Humeau H A.(2019)Texture Feature Extraction Methods: A Survey[J].IEEE Access,7:8975-9000.
[5] Park Y J, Guldmann J M. (2020) Measuring continuous landscape patterns with Gray-Level Co-Occurrence Matrix (GLCM) indices: An alternative to patch metrics?[J].Ecological Indicators,109:1-18.
[6] Akar Ö, Güngör O. (2015) Integrating multiple texture methods and NDVI to the Random Forest classification algorithm to detect tea and hazelnut plantation areas in northeast Turkey[J]. International Journal of Remote Sensing,36(2): 442-464.
[7] Wang H, Lu K, Pu R . (2016) Mapping Robinia pseudoacacia forest health in the Yellow River delta by using high-resolution IKONOS imagery and object-based image analysis[J]. Journal of Applied Remote Sensing, 10(4):045022.
[8] Smith J A (1980) The Lambertian assumption and Landsat data[J].Photogrammetric engineering & Remote sensing,46(9):1183-1189.
[9] Mitri G H, Gitas I Z . (2004) A semi-automated object-oriented model for burned area mapping in the Mediterranean region using Landsat-TM imagery[J]. International Journal of Wildland Fire, 13(3):367-376.
[10] Zhou J, Rui Y G, Sun M, et al. (2017) The Effects of GLCM parameters on LAI estimation using texture values from Quickbird Satellite Imagery[J]. Scientific Reports, 7(1):7366.

Gram-Schmidt Remote Sensing Image Fusion Algorithm Based on Matrix Elementary Transformation

Shuang Zhang[1*], Yifei Han[2], Hua Wang[1,2] and Daishuang Hou[1]

[1] School of Mechatronic Engineering, Changchun Institute of Technology, Changchun, 130012, China

[2] School of Mechatronic Engineering, Changchun University of Science and Technology, Changchun, 130012, China

[*] 416771138@qq.com

Abstract: In the application of the Internet of vehicles, it is difficult for a single remote sensing image to have accurate spatial information and spectral information at the same time, so engineers must use image fusion to improve the utilization of remote sensing image information. Aiming at the shortcomings of low spectral resolution and relatively complex programming of traditional fusion algorithms, this paper proposes a Gram-Schmidt remote sensing image fusion algorithm based on matrix elementary transformation, which aims to improve the spatial resolution of multispectral images by using panchromatic images with high spatial resolution. The orthogonalization process of the algorithm in this paper uses the elementary matrix transformation, which is simple and easy to perform. The experimental results show that it can effectively improve the spatial resolution of the original image and has good spectral retention.

1. Introduction

The automobile has made considerable progress since its birth with the continuous development of industrial technology, and with the development and maturity of the concept of the Internet of things, the Internet of vehicles has gradually entered people's lives. As an important part of the Internet of vehicles, remote sensing satellite plays an irreplaceable role in vehicle positioning, path planning, and road analysis. Engineers can analyze panchromatic images, multispectral images, hyperspectral images, and other remote sensing data captured by remote sensing satellites to obtain massive remote sensing information and apply it to the Internet of vehicles. But due to the restriction of spatial resolution and spectral resolution, a single original remote sensing image is very difficult to get accurate spatial information and spectral information, and realizes the high spatial resolution panchromatic image and preserves spectral information. However, relatively low spatial resolution multispectral image fusion has become the key to solving this problem.

At present, the commonly used remote sensing image fusion algorithms include the IHS algorithm, PCA algorithm, image fusion algorithm based on wavelet transform, Gram-Schmidt algorithm, et al. The IHS and PCA algorithms are both based on the digital matrix of the image for fusion processing. The difference is that the IHS algorithm converts the RGB component matrix of the image into I, H, and S components in the IHS space and finally achieves image fusion by replacing the I component and inverse transformation[6]. PCA algorithm is different from the IHS algorithm that can only handle three bands of disadvantage, PCA algorithm for multi-channel multi-spectral image and panchromatic image fusion, image matrix was first obtained after the decentralization of covariance matrix, then the eigenvalue and eigenvector, according to the size of the first principal component, the second principal component... PCA transformation and component replacement were performed on the obtained principal components, and finally, PCA inverse transformation was performed to obtain the fused image[10]. The wavelet transform is used to decompose the image, and the high-frequency detail and low-frequency detail components are obtained respectively[4]. Finally, the relevant fusion rules are used to process the final fusion model. The Gram-Schmidt algorithm (hereinafter referred to as the GS algorithm) makes an improved GS transform for each pixel and then obtains the fused image by component replacement and inverse transformation. Since the matrix will not change after GS orthogonalization and inverse transformation of the same matrix, the GS fusion algorithm has good spectral fidelity and can better improve the spatial details of the original image[11]. But the GS algorithm program design is more complex, and the steps are relatively tedious, the program implementation time is relatively long, so this paper is inspired by the GS transform, using the matrix of the elementary transformation of the GS transform. It has good spatial details and retains the spectral information of the original image to a large extent.

2. Introduction to the algorithm

2.1. Basic principle of GS orthogonalization

GS transformation is the most common way to normalize matrices in line generation, if there exist linearly independent vectors $\alpha_1, \alpha_2, \alpha_3 \ldots\ldots \alpha_n$ in Euclidean space V[8], then

$$\beta_1=\alpha_1 ; \tag{1}$$

$$\beta_2=\alpha_2-\frac{<\alpha_2, \beta_1>}{<\beta_1, \beta_1>}\beta_1 ; \tag{2}$$

$$\beta_3=\alpha_3-\frac{<\alpha_3, \beta_1>}{<\beta_1, \beta_1>}\beta_1-\frac{<\alpha_3, \beta_2>}{<\beta_2, \beta_2>}\beta_2 ; \tag{3}$$

$$\cdots \cdots$$

$$\beta_n=\alpha_n-\frac{<\alpha_n, \beta_1>}{<\beta_1, \beta_1>}\beta_1-\frac{<\alpha_n, \beta_2>}{<\beta_2, \beta_2>}\beta_2- \cdots - \frac{<\alpha_n, \beta_{n-1}>}{<\beta_{n-1}, \beta_{n-1}>}\beta_{n-1} ; \tag{4}$$

The $\beta_1, \beta_2, \beta_3, \ldots\ldots \beta_n$ are nonzero vectors, and orthogonal to each other, that is, for the α_s(S=1, 2, 3……n)results of GS orthogonalization, and then normalized to get the normalized orthogonal group.

2.2. GS algorithm

GS fusion algorithm was proposed by CraigA Laben et al., which eliminates redundant information by orthogonalizing the image matrix. The specific steps are as follows:

(1) Firstly, the panchromatic image with the low spatial resolution is simulated by using a multispectral image, namely, the first component of GS transformation GS_1. There are two different ways to simulate GS_1. The first method is to assign weights to different bands of a multispectral image to simulate a low-resolution panchromatic image, namely GS_1. The second idea is to blur the high-resolution panchromatic image, then take a subset of it, and finally resample it to the same size as the multispectral image to achieve image simulation.

(2) GS forward transformation is carried out using GS_1 obtained by simulation. The specific formula is as follows:

$$GS_T(i,j)=\left(B_T(i,j)-\mu_T\right)-\sum_{l=1}^{T-1} \varnothing(B_T,GS_l)\times GS_l(i,j) \tag{5}$$

where B_T is the T-th band of the original multispectral image, μ_T is the gray mean of the band, and $\varnothing(B_T, GS_l)$ is the covariance.

(3) Gray level matching and adjustment of panchromatic image and GS_1 are performed, and the

specific implementation method is shown in the reference.

(4) The adjusted panchromatic image is replaced with GS_1, and then the fused image is obtained by GS inverse transformation. The specific inverse transformation formula is as follows:

$$GS_T(i,j)=\left(B_T(i,j)+\mu_T\right)+\sum_{l=1}^{T-1}\varnothing(B_T,GS_l)\times GS_l(i,j) \tag{6}$$

2.3. GS fusion algorithm based on an elementary transformation

Based on the orthogonalization idea of the GS fusion algorithm, a new GS fusion algorithm based on matrix elementary transformation is proposed[2]. The algorithm flow is as follows:

(1) By transforming the multi-spectral image with size m*n*3 into digital matrix A ($t=m\times n$) of t*3, the real symmetric matrix $Q=A^TA$ and the augmented matrix $K=(A^TA, A^T)$ can be obtained.

(2) Perform elementary row transformation on the augmented matrix K, and finally get the upper triangular matrix M of Q and the matrix B(3*t) of A^T after the same elementary row transformation. Matrix B is the orthogonalization result of A^T.

(3) The first row of B is transformed into the image matrix I_1 of m*n, and the panchromatic image P and I_1 are gray matched to obtain a new P.

(4) The modified panchromatic image newP is replaced with I_1, and the fused image F is obtained by inverse transformation of orthogonal processing. The specific formula is as follows:

$$F_3=B_3+B_2*\alpha_1+B_1*\alpha_2 \tag{7}$$

$$F_2=B_2+B_1*\alpha_3 \tag{8}$$

where α_1, α_2 and α_3 are the coefficients produced in the process of orthogonalization.

3. Analysis of experimental results

3.1. The experimental results

To evaluate the effect of the proposed algorithm, two groups of different types of data are used for fusion experiments. Data 1 Remote sensing data: a 30M resolution multispectral image named (a) and a 15M resolution panchromatic image named (b) of a region in Qingdao acquired by Landset8 satellite in February 2021 were selected, and the relevant three-color band fusion was carried out. Besides, data 2 Remote sensing data includes a four-band multispectral image named (b) with a resolution of 3.2m and a panchromatic image named (b) with a resolution of 0.8m taken by GF-2 satellite in May 2019 in a certain place in Kuancheng District, and the three-band multispectral image was obtained by removing the near-infrared band. All the data were preprocessed by strict geometric correction, ortho-correction, image registration, and resampling, and the same ROI region was clipped and compared with the fused images obtained by the IHS algorithm, PCA algorithm and GS algorithm. The images before and after fusion are shown in the figure below.

(a) MS (b) PAN (c) IHS algorithm

(d) PCA algorithm (e) GS algorithm (f) The algorithm in this paper

Figure 1. Test the effect of different fusion algorithms in data 1.

(a) MS (b) PAN (c) IHS algorithm

(d) PCA algorithm (e) GS algorithm (f) The algorithm in this paper

Figure 2. Test the effect of different fusion algorithms in data 2.

3.2. Quality evaluation

The quality evaluation is an indispensable key step in remote sensing image fusion. The visual evaluation results are obtained by testing and comparing the images before and after fusion as well as various fusion images. Through the analysis of the evaluation results, the deficiency of the algorithm can be reflected, and the algorithm can be optimized by adjusting accordingly.

3.3. Intuitive evaluation

Firstly, intuitively, several fusion methods improve the spatial resolution of remote sensing images and maintain the original spectral information in different degrees. Secondly, the horizontal comparison of several fusion methods shows that the texture details of the four methods have been enhanced to different degrees, and different ground object information can be identified. In terms of spectrum preservation, it can be seen that the IHS algorithm has a large spectral distortion in the water and vegetation areas, and the PCA algorithm also has a certain degree of spectral loss in the building areas. GS and the proposed algorithm both maintain the original spectral information well[1].

3.4. Objective evaluation

To obtain more accurate evaluation criteria, the remote sensing images were quantitatively evaluated from the aspects of gray mean, information entropy, mean gradient, spatial frequency, standard deviation, signal-to-noise ratio and correlation coefficient[9]. As shown in the following table:

Table 1. Effects of different fusion algorithms in test data 1.

	Average	Information entropy	Average gradient	Spatial frequency	Standard deviation	SNR	Correlation coefficient
PAN	43.5188	6.3226	6.6992	22.1685			
MS	52.2358	6.4988	3.592	10.7325			
IHS	54.1353	6.5691	6.356	21.237	0.0158	14.4264	0.8824
PCA	52.3277	6.4730	5.9462	19.8045	0.0152	12.7155	0.9286
GS	52.2546	6.4977	5.8169	18.7321	0.0107	11.1405	0.9456
The algorithm in this paper	52.3278	6.7184	6.9565	23.0031	0.0145	15.1304	0.9306

Table 2. Effects of different fusion algorithms in test data 2.

	Average	Information entropy	Average gradient	Spatial frequency	Standard deviation	SNR	Correlation coefficient
PAN	52.386	6.6063	6.7323	18.8214			
MS	63.3806	7.1527	2.2431	4.9326			
IHS	63.5717	6.8624	6.2777	16.8535	0.0227	35.4831	0.8041
PCA	63.4085	7.2015	7.7749	21.2765	0.0284	44.3424	0.72
GS	63.4533	7.0125	7.3764	19.7787	0.0258	40.29	0.7521
The algorithm in this paper	63.2492	7.2565	7.8849	21.6857	0.0239	37.3716	0.8085

It can be concluded from the table that the fusion image obtained by the proposed algorithm has the best comprehensive performance in each evaluation index. The IHS algorithm and PCA algorithm are smaller than the proposed algorithm in terms of information entropy, average gradient, spatial frequency and correlation coefficient, and the vegetation and water area are seriously distorted. Compared with the GS algorithm, the comprehensive evaluation index of the proposed algorithm is superior. From the chart, it can be objectively reflected that the proposed algorithm has obvious advantages compared with other algorithms in the improvement of spatial resolution and the maintenance of spectral information.

4. Conclusions

This paper is based on the matrix operation in the image fusion method, inspired by the GS fusion algorithm, this paper proposes a GS remote sensing image fusion algorithm based on matrix elementary transformation: first of all, the two-dimensional matrix elementary row transformation gets the upper triangular matrix, the resulting matrix orthogonalization results, and then replaces image which is replaced with the image grayscale processing, and completes the corresponding replacement, finally uses the inverse transformation related to the final fusion image. The experimental results show that this method is relatively simple in program design and has good spectral retention while improving the spatial resolution of the original image.

Acknowledgments

This research was supported by the Jilin province science and technology development-funding project, and the project serial number are 20210203037SF, 20210502025ZP and 20210101486JC. Meanwhile, it was supported by Capital construction funds within the budget of Jilin Province in 2021 (innovation capacity building), and the project serial number is 2021C044-6.

References

[1] Wang Y, YU L X, Wang C X, Zuo F and Wang Z H 2021 Remote Sensing Image Fusion Algorithm Based on Improved IHS, PCA and Wavelet Transform. *Computer & Digital Engineering,* 49(04): 797-803.

[2] He Z K, Zhu Y Z and Liu Q X 2021 A Schmidt Orthogonalization Method Based on Elementary Transformations. *College Mathematics*, 37(02): 42-47.

[3] Liu C, Qi X D, Zang W Q, Huang X Z and Yang X F 2018 Research of improved Gram Schmidt image fusion algorithm based on IHS transform. *Engineering of Surveying and Mapping*, 27(11): 9-14.

[4] Wang H 2019 The Study on Hyperspectral Remote Sensing Image Fusion Based on Wavelet Transform. *Nanchang Hangkong University*, Nanchang.

[5] Ma Y W 2010 Research on Hyperspectral Remote Sensing Image Fusion and Quality Evaluation Method. *PLA Information Engineering University*, Zhengzhou.

[6] Dai J 2020 Application Research on IHS Fusion Method of Multi-source Remote Sensing Image. *Changsha University of Science & Technology*, Changsha.

[7] Yin X Y and Ye Y S. 2020 Comparison and Evaluation of Remote Sensing Image Fusion Methods. *Sci-tech Innovation and Productivity*, (06): 52-56+60.

[8] Ge C R 2019 Research on Hyperspectral Image Based Multi-source Remote Sensing Image Fusion. *Xidian University*, Xian.

[9] Tang M 2019 Three remote sensing image fusion and evaluation based on Matlab. *Journal of Heilongjiang Institute of Technology*, 33(05): 40-44.

[10] Li C J, Liu L Y, Wang J H and Wang R C 2004 Comparison of Two Methods of Fusing Remote Sensing Images with Fidelity of Spectral Information. *Journal of Image and Graphics*, (11): 106-115+128-129.

[11] Yu H Y, Yan B K, Gan F P, Chi W X and Wu F D 2007 Hyperspectral Image Fusion by an Enhanced Gram-Schmidt Spectral Transformation. *Geography and Geo-Information Science*, (05): 39-42.

Extraction Method of Stand Density Based on High-Resolution Remote Sensing Imagery

Xuefeng Han[1a*], Huan Xu[1b*]

[1]Department of Resource Exploration, TianJin Petroleum Vocational and Technical College, Tianjin 301607, China

[a*]hanxf6@163.com, [b*]13920699972@163.com

Abstract— In this paper, a stand density extraction method is proposed based on QuickBird high-resolution satellite image data of Baisha artificial forest farm in Minhou, Fujian. Firstly, through the analysis band canopy image of QuickBird panchromatic, it is known that there is a maximum reflectivity point in the canopy. Then, the vegetation-covered area was extracted by the normalized difference vegetation index (NDVI). The canopy image of the forest was extracted through the decision tree classification algorithm. The texture information and spectral information were comprehensively utilized to classify the images. Then, the number of stand trees was extracted by the maximum filtering method, and the stand density was calculated to produce the stand density map. The experiment verified that this method can meet the needs of a stand density survey.

1. Introduction

Forest ecosystems harbor a large proportion of global biodiversity, contribute extensively to biogeochemical cycles, and provide countless ecosystem services, including water quality control, timber stocks, and carbon sequestration. Tree density is a prominent component of ecosystem structure [1-4]. Forest number density refers to the number of trees per unit area. It is one of the important indicators of forest resource survey and plays an important role in indicating biodiversity, forest volume, forest structure, etc[5]. In the past, high precision stand density was obtained mainly through artificial field investigation. Although the accuracy was high, the labor intensity was large and the investigation range was very limited. Remote sensing data can be used for forest resource surveys. TM, SPOT and other low-and medium-resolution remote sensing data can be used to estimate the characteristic parameters of forests greater than 100 hectares[6]. The forest number density at the regional scale was mainly estimated by combining the ground survey data. There had been research results on global-scale forest plant density mapping[3], but its resolution was low and lacks detailed spatial information. Fine stand density is often based on field survey data or high-resolution remote sensing data[6]. Most individual tree identification studies have focused on the use of digitized aerial photographs or airborne data. However, the disadvantages of these data include inherent geometric artefacts that were a function of the camera or sensor optics, coupled with the relatively small area of the ground they typically cover. High spatial resolution satellite sensor imagery, as described by Aplin et al. (1997), has the potential to negate some of the common problems associated with airborne data and to facilitate the identification of detailed structural attributes of forest cover over large areas[7]. With the rapid development of remote sensing technology, the resolution of remote sensing images has been greatly improved, such as the panchromatic band of the QuickBird satellite image, the spatial resolution has reached 0.61m, which is equivalent to the aerial image with a scale of about 1: 5000.

Taking QuciBird image as an example, this paper explored the method of comprehensively using spectral information and texture information in high-resolution remote sensing image, and then extracting the number of forest trees by maximum value filtering method, and then calculating the forest density.

2. Study site and data acquisition

The Study site in this paper is located in Baisha Forest Farm in Minhou, Fuzhou, in the eastern part of Fujian Province. It is mainly distributed in the middle and low mountains, belonging to the subtropical marine monsoon climate. The winter is short and summer is long, and the frost-free period is 326 days. The annual average sunshine duration is 1700-1980hours, and the annual average precipitation is 900-2100 mm. The annual average temperature is 19.6°C, and the average temperature is 77%. The climate is mild, and the rainfall is abundant. The four seasons are evergreen. The management area of the forest farm is about 4139.2hm^2, of which the volume of standing trees is 698 000m^3, and the coverage rate of forest grazing reaches 94.9%. The stand grows well. Trees in forest farms were mainly fast-growing timber forests.

Details of QuickBird data were included in table1. A 0.61m panchromatic QuickBird image was acquired in 2016. The QuickBird image was orthorectified and spatially registered to the MEIS II data and stem map. The QuickBird image was orthorectified (Figure.1) to eliminate the image deformation with DEM.

The validation data mainly includes the forest stand survey data and a small class vector map provided by Minhou Baisha Forest Farm in Fuzhou.

Figure.1 QuickBird imagery after ortho-rectification

3. Extraction Method of Stand Density

3.1 Spectral reflection characteristics of crown

Chen Jingming (1997) simulated spectral reflectance characteristics of crowns at different zenith angles of solar rays. At different solar elevation angles, the crown has different reflectivity characteristics, but all have a maximum reflectivity point. According to the shooting time of high-resolution remote sensing data, the solar zenith angle of the QuickBird image is calculated to be 36.08 degrees. In this paper, the real spectral reflectance characteristics of coniferous and broad-leaved crowns are studied (Figure.2). It can be seen from the graph that there is a maximum spectral reflection point on the crown surface, especially on the coniferous crown. Since the canopy has the maximum reflectivity point, the number of canopies can be extracted by extracting the maximum reflectivity points of canopies in a high-density forest, and then the stand density can be calculated.

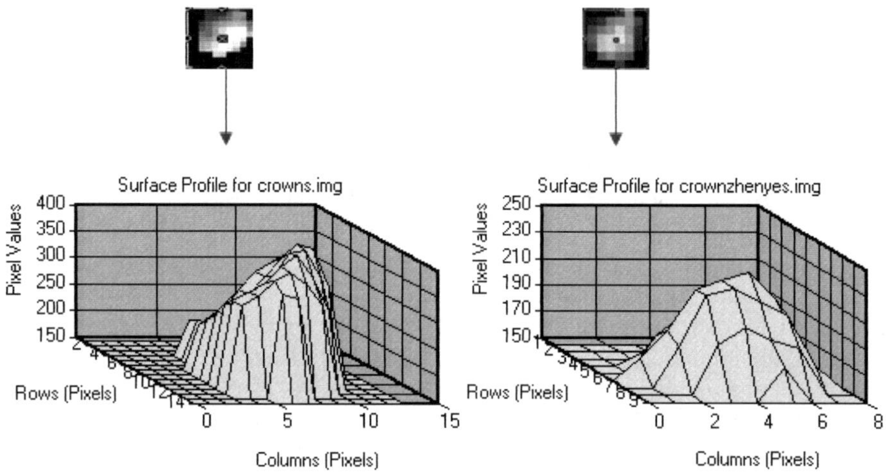

(a) (b)

Figure.2 Spectral reflectance distributions of the canopies(a: conifer canopy b: broad-leaved tree canopy)

3.2 Extraction of the forest stand density.

First, the orthorectification of this high-resolution remote sensing image was carried out, and the panchromatic band with a resolution of 0.61m and the multispectral band with a resolution of 2.44m were fused. Then, the normalized difference vegetation index (NDVI) and texture information were calculated based on the fused image, and the NDVI was used to extract the vegetation coverage area. Next, the decision tree classification algorithm was used to classify the vegetation index NDVI and texture information to extract the forest canopy image. Afterward, the local maximum filtering was carried out to obtain the maximum reflectance points from QuickBird panchromatic band image, and the stand density was calculated. Finally, the stand density map was made based on the stand planning map.

The setting of window size is important in the algorithm of local maximum filtering. The selection of optimal window size is key[8]. When the window size is too small, nonexistent trees were detected because an individual tree crown is divided into several parts. If the window size is too large, existent trees are missed due to the window containing several trees[9]. In this paper, through continuous experiments, it was found that the number of stand trees was closest to the actual survey data when the window size was set to be equivalent to the crown size.

4. Stand density extraction experiment

The maximum reflectivity points of canopies were extracted from the canopy image, so the canopy image was first segmented from the image. This process includes two steps. Firstly, the NDVI is calculated from the spectral information of the image, and vegetation and non-vegetation areas were classified to extract vegetation cover area. Then, the vegetation classification method was used to extract the forest canopy imagery by the comprehensive application of the cultural and spectral information of the imagery.Step1:Extraction of vegetation-covered area. Studies have shown that normalized difference vegetation index (NDVI) can accurately extract vegetation cover area from high-resolution images[10]. The NDVI was defined as the ratio of the difference between the near-infrared band and the visible red band to the sum of the two band values. According to the statistics of various vegetation indexes in the study area (Table 1), the NDVI values in the vegetation cover area were generally greater than 0.281, and the NDVI values in the bare land and shadow areas were generally less than 0.281. Step2: Extraction of forest canopy imagery. The vegetation coverage area in the study area includes the grassland coverage area and canopy coverage area. It was difficult to segment canopy images from spectral information of vegetation. The texture features of forest and grassland in high-resolution images were different. To accurately extract the image of the canopy cover area from the vegetation cover area, the texture information of the image was also needed. The symbiotic matrix is an efficient statistical method for texture analysis. The local consistency index (Homogeneity) was used to identify different vegetation types. The local consistency index is calculated by the following formula. Decision tree classification has better flexibility and robustness for spatial features and classification marks of input data[6]. Firstly, the typical sample plots of woodland and grassland were selected from the study area through visual interpretation as training samples. Then, a decision tree model was established by using the decision tree algorithm C5.0, and the classification rules were extracted. Finally, the classification tree was pruned according to the reliability to obtain a classification decision tree with high reliability. Finally, the decision tree module in ENVI4.7 was used to automatically classify images. The study area was divided into woodland and grassland by decision tree classification. The confusion matrix was obtained by the test method of the ENVI region of interest, and its accuracy was evaluated (Table 2). The overall accuracy of classification was 91.933%, and the Kappa coefficient was 0.8043. The classification accuracy was high, which could meet the needs of the forest resource survey.

Table 1. Statistical values of NDVI of different land cover types

Land Cover Types	Maximum	Minimum	Mean
Grassland	0.430	0.321	0.380
Broad-leaved forest land	0.593	0.462	0.517
Coniferous forest land	0.507	0.319	0.413
Bare land	0.281	0.123	0.196
Tree shadow coverage	0.250	0.229	0.239

$$f_{Hom} = \sum_{N-1}^{N-1}\sum_{i=0}^{N-1}\frac{P(i,j)}{1+(i\cdot j)^2}$$

Where N is the order of the symbiosis matrix, i, j are the coordinates of the symbiosis matrix, and P is the value of the symbiosis matrix at (i, j).

Table 2. Vegetation classification accuracy

types	grassland	forest land	sum	User's Accuracy/%
grassland	91852	21426	113278	81.09
forest land	8290	246798	255088	96.75
sum	100142	268224	368366	¾
Producer's Accuracy/%	91.72	92.01	¾	¾
Overall Accuracy = 91.933%		Kappa Coefficient = 0.8043		

Stand density is extracted. Forest canopy vertices were obtained by extracting the point of maximum crown reflectivity in the study area. First, the average crown size was estimated in different regions. Then, according to the estimation results, the filter window which was equivalent to the crown size was set. Through repeated tests, the best filter window was set to 3 * 3. Then, the 3 * 3 maximum filtering was performed on the QuickBird panchromatic band to obtain a local maximum filtering grid layer. Then, the difference between the filtered layer and the original panchromatic band layer was calculated. The difference was zero, which was the maximum reflection point of the crown. Then, the extracted maximum reflection point of the crown was superimposed with the NDVI layer to extract the NDVI value of these vertices. Finally, a certain threshold was set to remove the vertices that were less than a certain NDVI value. Finally, more accurate maximum reflection points of the crown were obtained (the red cross point in Figure. 3 represented the maximum reflection point of the extracted crown).

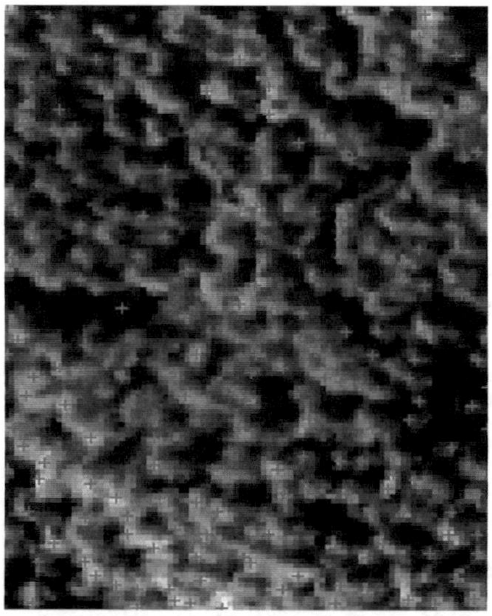

Figure.3 Tree Identification by high-resolution remote sensing

(2) Evaluation of the extraction accuracy

In this paper, the average density in the three types of survey data of forest farms in the study area (some survey data of cutting area) was used as the real value to test the tree crown extraction algorithm. The results were shown in Table 3.

Table 3. Precision test of forest density extracted from remote sensing

Class number	The density of the survey (Plant / acre)	The density of estimating (Plant/acre)	Opposite error	Percentage error(%)
402010	153	137	16	11.7
402120	199	185	14	7.6
403060	261	243	18	7.4
403070	209	199	10	5.0
405060	208	229	21	9.2
406080	240	225	15	6.7
407010	178	162	16	9.9

From Table 3 and Figure 3, it can be seen that the accuracy of the maximum point extraction of canopy reflectance was relatively high, the maximum error was11.7%, and the minimum error was 5.0%, which can better reflect the density of dominant trees. However, at the same time, the estimated density was smaller than the survey density. There were three main reasons. First, the image presents only the upper canopy of trees, and some short crowns were covered by adjacent higher trees and cannot be identified from the image. Second, because the filter window was too large, the maximum canopy reflection point of the two closer trees was included in a filter window range, then only one can be extracted, resulting in a small extraction result.

5. Conclusions

High-resolution remote sensing images make it possible to extract individual tree information. This paper mainly discussed the extraction method of forest stand density based on high-resolution remote sensing images. This paper proposed a method to extract forest density by using a decision tree algorithm to extract forest canopy images and using a local maximum filtering algorithm to extract the maximum point of canopy reflectance. The method used a decision tree image classification algorithm to extract forest canopy images, eliminated background information, avoided extracting individual tree information in a non-crown area, and improved the accuracy of forest density extraction. The application of this method has a premise assumption: under certain conditions, the reflectivity of the crown vertex was maximum. This assumption does not apply under all conditions. The precision is high for mature forests, but low for the young forest, high for the coniferous forest, but low for the broad-leaved forest. In terms of filtering technology, the window size was a key factor in its density estimation. Transforming filtering windows according to different canopy sizes was better than using the same window. In this sense, the accuracy of stand uniformity was higher, and the accuracy was lower when the crown width changes greatly. Therefore, how to automatically transform the filtering window according to the image situation was the focus of future research.

References

[1] Crowther, T. W. et al. Predicting the responsiveness of soil biodiversity to deforestation: a cross-biome study. Glob. Change Biol. 20, 2983–2994 (2014).
[2] Hansen, M. C. et al. High-resolution global maps of 21st-century forest cover change. Science 342, 850–853 (2013).
[3] Crowther T W et al. Mapping tree density at a global scale.[J]. Nature, 2015, 525(7568) : 201-5.
[4] Suhan wang et al. Stand density estimates based on a local maximum spectral filter[J].journal of Zhejiang a & f university,2017,34(03):413-420.
[5] GUO Wei, YANG Chunyu, et al. Forest density mapping based on UAV high—resolution image[J].Bulletin of Surveying and Mapping, 2021(07):29-33. DOI:10. 13474/j.cnki. 11-2246.2021.0204.
[6] Hyyppä J, Mielonen T, Hyyppä H, et al. Using individual tree crown approach for forest volume extraction with aerial images and laser point clouds[C]//Proceedings of The ISPRS Workshop Laser Scanning 2005. 2005: 144-149.
[7] images and laser point clouds. ISPRS WG III/3, III/4, V/3 Workshop "Laser scanning 2005", Enschede, the Netherlands, September 12-14, 2005.
[8] Midhun Mohan,Bruno Araujo Furtado de Mendonça,Carlos Alberto Silva,Carine Klauberg,Acauã Santos de Saboya Ribeiro,Emanuel José Gomes de Araújo,Marco Antonio Monte,Adrián Cardil. Optimizing individual tree detection accuracy and measuring forest uniformity in coconut (Cocos nucifera L.) plantations using airborne laser scanning[J]. Ecological Modelling,2019,409(C).
[9] Xu X, Zhou Z, Tang Y, et al. Individual tree crown detection from high spatial resolution imagery using a revised local maximum filtering[J]. Remote Sensing of Environment, 2021, 258: 112397.
[10] CHEN Junying,TIAN Qingjiu.Vegetation Classification Based on High-resolution Satellite Image[J]. Journal of Remote Sensing,2007(02):221-227.

Research on the Application of Small Area 3D Modeling Based on Consumer UAV—Take DJI Yu-2 UAV as an Example

Zhan Teng[1, a], Wanqing Liu[2, b], Yaowu Yi[2, c]

[1]Northwestern University, Shaanxi, Xian, 71000012, China

[2]Air Force Engineering University, Shaanxi, Xian, 7100003, China

[a]*Corresponding author's e-mail: gyx84562@163.com

[b]Author's e-mail: O123456HUV@163.com

[c]Author's e-mail: O123456KMT@163.com

Abstract: The use of UAVs for geographic information mapping is a relatively mature technology. It has the advantages of a large measurement range, fast data collection speed and high efficiency. It is gradually becoming an important technical means that can partially replace traditional field measurement. 3D modeling of UAV image data is an important aspect of UAV surveying and mapping applications. Under the current background of "real China" and "digital twin", UAV modeling will have very broad application prospects. The current UAV modeling is mainly based on large fixed-wing UAVs or professional rotary-wing UAVs, but their prices are relatively high, so this paper tries to use a relatively low-priced consumer-grade UAV-DJI Yu 2, the professional version of the UAV performs image control-free 3D modeling without ground control points and analyzes the accuracy of the results to verify its applicability in geographic information data collection.

1. Introduction

UAV has become the main method tool for spatial data acquisition and engineering surveying and mapping. Compared with traditional surveying technology, UAV surveying and mapping has obvious advantages in terms of operation efficiency, data accuracy, and economic cost. With the development of drones in recent decades, it has experienced the development of film images to digital images, and mapping technology has also developed from analog stereo relative measurement to the current interactive/automatic digital data collection. UAV hardware technology has also developed from manual operation to semi-automatic operation, and then to intelligent flight. At present, the most widely used UAVs in the market and industry mainly include fixed-wing UAVs and rotary-wing UAVs.

The fixed-wing UAV is derived from manned aircraft. It is characterized by its large size, long flight time and mileage, and large carrying capacity. It can carry large-scale photography equipment and remote sensing equipment such as multi-spectral cameras; it has high-precision GNSS positioning. and POSE sensors, which can obtain high-precision image external orientation elements, and are generally used in large-scale engineering surveying and mapping applications with high precision requirements. The generation and application of rotary-wing UAVs in space surveying and mapping is relatively short. It has the characteristics of low site requirements, flexible maneuverability, and low economic cost. It is gradually becoming the main productivity of low-altitude, medium/small area surveying and mapping. Rotor drones can be divided into professional drones and consumer drones according to flight distance, take-off weight, and functional modules.

Here, the campus of Shaanxi Vocational and Technical College of Transportation is used as the project survey area, and the basic method and process of photogrammetric modeling using consumer drones are briefly described [8].

2. Data acquisition and processing

2.1 Technical process of UAV surveying and mapping

The basic technical process of traditional UAV surveying and mapping includes: ① laying out image control points ② route planning ③ shooting images ④ image registration ⑤ building stereoscopic images ⑥ data collection, generation of DEM, DOM, DSM and other products. The traditional UAV surveying and mapping process is relatively complex and requires a lot of manpower and material resources. It is widely used in engineering projects with high precision requirements. With the addition of satellite positioning technology (GNSS) and attitude control system (POSE) to the UAV photogrammetry system, images with high-precision positioning information and aerial attitude information can be directly obtained, thereby eliminating the need for ground control measurement. The work efficiency is further improved. This modeling adopts the image-free method.

2.2 Planning for this mission

2.2.1 Introduction to UAV

The UAV used this time is the DJI Yu-2 UAV (referred to as “Yu-2” UAV), which is one of the most popular UAV products on the market. It has many advantages such as high reliability, high intelligence, high imaging quality and relatively low economic cost. It is the best choice for surveying and mapping in consumer drones.

2.2.2 Mission planning

The flight control software used for this flight is DJI PILOT. Because the Royal 2 is a consumer drone for general users, there is currently no flight control software for surveying and mapping engineering for the Royal 2 drone, but it can Share this APP with DJI Royal series industrial drones for mission planning, route setting and other flight control operations.

The main flight parameters are set as follows:

① Overlap rate: side (tilt) -60%, heading (tilt) -70%; side (ortho) -70%, heading (ortho) -80%

② Route flight height: In order to ensure flight safety, the flight height should be higher than the tallest building in the survey area. If there is a building with an uncertain height, you can first drive the drone to approach it, adjust the gimbal to a horizontal state, and the center of the lens coincides with the top of the building. The current altitude of the drone can be used to estimate the building height. Increase the flight height by 10-20 meters, which can be used as the minimum flight height for safe flight. In addition, the flight height will affect the number of images and the final model accuracy: the flight height is negatively correlated with the number of images and model quality, and has a positive correlation with the operation efficiency, that is, the lower the flight height, the more images and the higher the model quality. The longer the work time [9].

③ Route flight speed: The faster the flight speed, the higher the operation efficiency, but it is easy to cause factors that reduce the image quality such as picture blur and jelly effect, so the default value of 5 m/s is used here.

④ Pitch angle of the gimbal: The pitch angle of the gimbal needs to be determined according to the height of the ground object, the height difference of the undulating height, and the height of the flight. It is set to 60° here.

⑤ Shooting mode: There are two modes of shooting mode: equal time interval and equal distance interval. In the area with flat terrain, both modes are available; when the terrain is undulating or there are high-rise objects, the equal distance mode can be used. to ensure uniform image sampling. Shot in equal time interval mode here.

2.2.3 Image data

During this flight, a total of 543 images were obtained, as shown in Figure 1.

Figure 1. Image data list

2.2.4 Aerial triangulation

The software used for model reconstruction is CONTEXT CAPTURE 4.3, the computer configuration is Intel CORE i5, and the memory is 8GB. The total time is about 10 hours. Figure 2 shows the point cloud results from aerial triangulation.

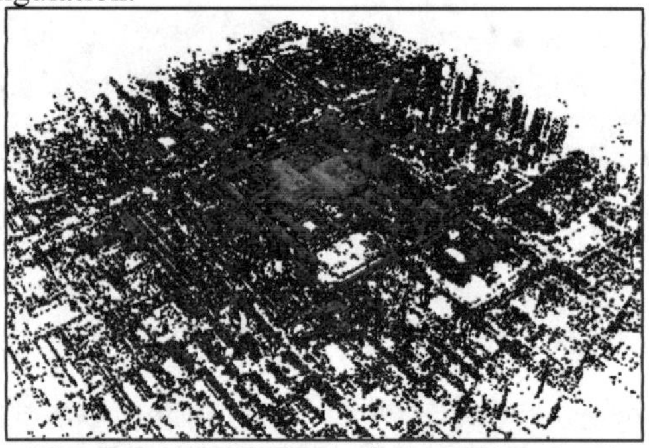

Figure 2. Aerial triangulation point cloud

3. Case analysis - 3D modeling of Shaanxi Jiaotong Institute

3.1 Survey area overview

(1) Area of survey area: The area is about 370m*410m≈0.15 square kilometers.

(2) Operation time: There are 5 routes in total, and the total time is about 55 minutes.

(3) Modeling process: The basic steps for reconstructing the model are as follows: ①Importing images ②Aerial triangulation operations ③Dividing cells according to the amount of data ④ Generating the model

(4) Model results: The software used for model reconstruction is CONTEXT CAPTURE 4.3, the computer is configured as Intel CORE i5, and the memory is 8GB. It takes about 10 hours. The results are shown in Figure 3.

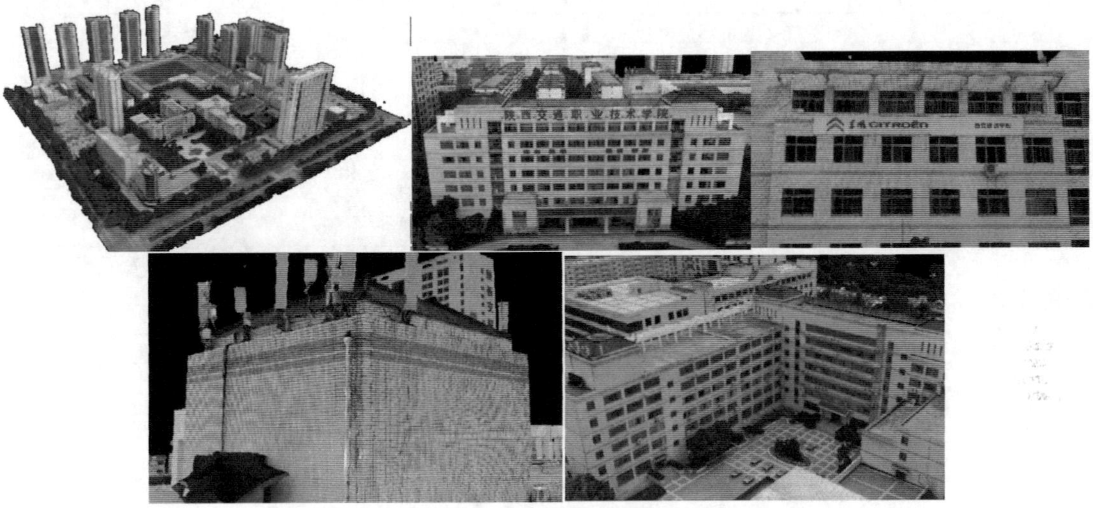

Figure 3. 3D model

3.2 Precision analysis

Select evenly in the survey area: roads, buildings, municipal facilities, et al., measure the length, width, height, height difference, et al. on the model, and compare with the topographic map measured by the total station to evaluate the model accuracy. Figure 4 shows the selected features.

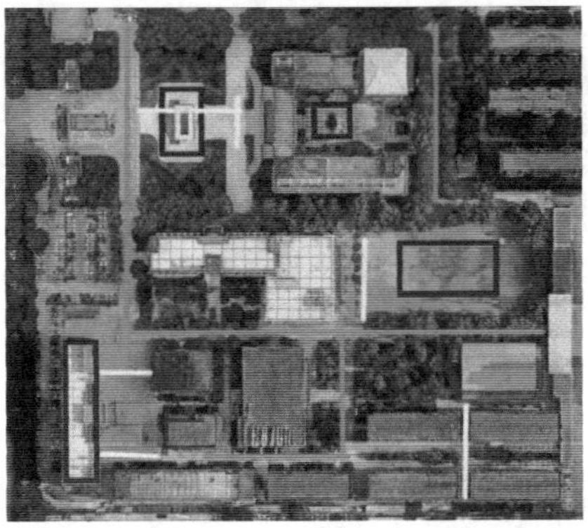

Figure 4. Select verification features

The points, lines, and surface features selected above are compared with the measured data and summarized in Table 1. The relative error of most of the geometric dimensions is 0.5-0.9%, the relative error of the area of the rectangle is 1.5-2.5%, and the area of one rectangle is the roof area. Due to its high height, the relative error is small, 0.75%.

Table 1: Comparison table of geometric dimensions of ground objects

serial number	features	Model size	Measured size	Relative error
1	the way 1	165.93	164.5	0.87%
2	the way 2	97.28	96.58	0.72%
3	the way 3	43.11	42.71	0.94%
4	the way 4	223.76	222.3	0.66%
5	the way 5	101.21	100.49	0.72%
6	basketball court center-line 1	15.13	15.05	0.53%
7	basketball court center-line 2	15.17	15.04	0.86%
8	basketball court center-line 3	15.11	15.01	0.67%
9	basketball court center-line 4	15.11	15.02	0.60%
10	basketball court center-line 5	15.17	15.07	0.66%
11	basketball court center-line 1	28.3	28.07	0.82%
12	basketball court center-line 2	14.14	14.02	0.86%
13	distance between two points 1	26.77	26.62	0.56%
14	distance between two points 2	33.86	33.63	0.68%
15	distance between two points 3	43.11	42.71	0.94%
16	distance between two points 4	34.28	34.15	0.38%
17	distance between two points 5	56.55	56.13	0.75%
18	distance between two points 6	25.79	25.56	0.90%
19	Rectangle 1	63	61.5	2.44%
20	Rectangle 2	849.39	837.2	1.46%
21	Rectangle 3	681.15	671	1.51%
22	Rectangle 4	2528	2487.9	1.61%
23	Rectangle 5	1508	1496.7	0.75%

3.3 Model application

On the model, coordinate measurement, spatial geometry measurement, line-of-sight analysis, and further generation of DEM and DOM products can be performed. The DEM and DOM generated according to the model are shown in Figure 5.

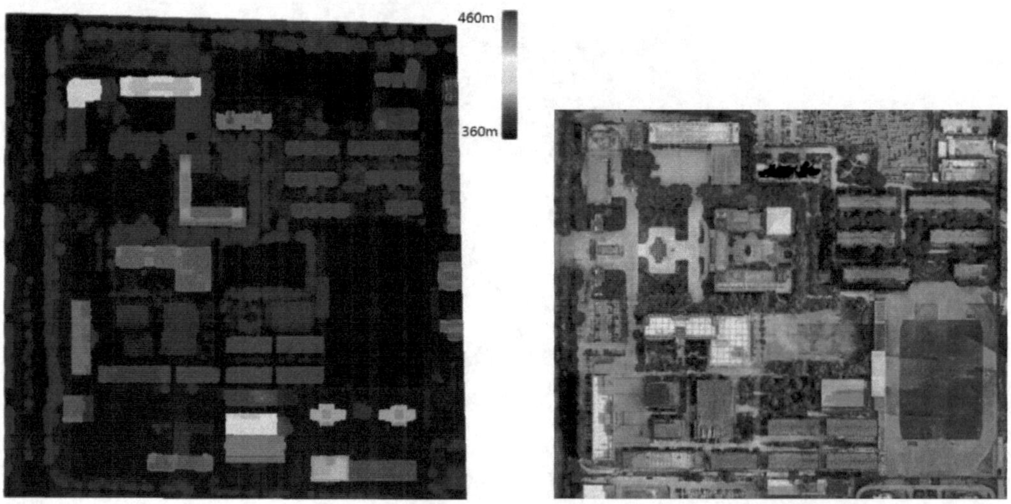

Figure 5. DOM and DEM

4. Conclusion

Through this modeling example, it can be determined that consumer-grade UAVs are feasible in photogrammetry and 3D modeling, the technical threshold is low, and ordinary users can perform surveying and mapping operations after a certain amount of learning and training. Its economic cost is currently the lowest standard, and the relevant supporting computer equipment only needs a general configuration, which can meet the economic feasibility of individual UAV photogrammetry and mapping in small areas. Observing the model, it can be found that the current modeling methods and results still have several obvious shortcomings: ① In the case of occlusion (such as trees, et al.), the ground objects are prone to obvious deformation and distortion, and the vegetation texture is difficult to accurately represent; ② The spatial size accuracy of the model is not uniform enough, for example, the length accuracy differs from the area accuracy by about 2-3 times. In order to make up for the above deficiencies, when planning the mission, it is necessary lowering the flight height, and taking supplementary shots to consider adding manual flight operations, for subjects that are difficult to shoot at high altitudes but pay attention to the uniform shooting interval and angle.

Acknowledgment
Project No.: 2021 General Special Scientific Research Plan of Shaanxi Provincial Department of Education

References

[1] Chen Xiaoyong, He Haiqing, Zhou Junchao, et al. Status and prospect of stereo image matching for low-altitude photogrammetry [J]. Chinese Journal of Surveying and Mapping, 2019, 48(12): 1595-1603. DOI: 10.11947/j. AGCS. 2019.20190466.
[2] Liu Jiancheng, Wang Guanzhi, Jin Zelin, et al. Three-dimensional modeling and accuracy analysis of oblique photogrammetry for urban real scenes [J]. Bulletin of Surveying and Mapping, 2021(S1): 16-19.DOI: 10.13474/j.CNKI.11- 2246.2021.0503.
[3] Li Tian. Fine 3D modeling of tower building based on wrap-around oblique photography [J]. Bulletin of Surveying and Mapping, 2019(S1): 318-321. DOI: 10.13474/j. CNKI. 11-2246.2019.0574.
[4] Sun Baoyan, Wei Longhua, Zhou Xin, et al. 3D modeling of abrupt and large height difference sites based on consumer drones [J]. Bulletin of Surveying and Mapping, 2021(7):111-116. DOI: 10.13474/j. CNKI. 11-2246.2021.0219.
[5] Tan Rinchun, Li Pengpeng, Wen Lin, et al. Optimization of urban 3D modeling method for UAV oblique photography [J]. Bulletin of Surveying and Mapping, 2016(11):39-42. DOI: 10.13474/j. CNKI. 11-2246.2016.0361.
[6] Wang Guo, Hu Qingfeng, Gong Tianhao, et al. Interactive building mapping method for oblique photographic models [J]. Bulletin of Surveying and Mapping, 2021(S1): 308-3 11.DOI: 10.13474/j.CNKI.11-2246.2021 .0570.
[7] Wen Caihuan, Wang Wendong, Wang Xiaodong. Layout and analysis of image control po ints for spatial information acquisition of real 3D model [J]. Surveying and Mapping Bulletin, 2021(S1): 1-5.DOI: 10.13474/j.CNKI.11-2246.2021. 0500.
[8] Ye Weilin, Pleiades, Wei Wanhong, et al. Application of UAV aerial survey system in la ndslide emergency [J]. Bulletin of Surveying and Mapping, 2017(9):70-74. DOI: 10.1 3474/j. CNKI. 11-2246.2017.0290.
[9] Zhao Yizhan, Zhou Lu, Pan Yuanjin, et al. Verification of tree barrier monitoring accurac y for general consumer UAVs [J]. Bulletin of Surveying and Mapping, 2021(S1): 159 -164.DOI: 10.13474/j.CNKI. 11-2246.2021.0536.

Analysis of Integrated Navigation Technology Based on GNSS/UWB

Ye Tian, Lejie Zhou, Xu Ji

School of Geomatics and Urban Spatial Informatics, Beijing University of Civil Engineering and Architecture, Beijing 102616, China

E-mail: 757254061@qq.com

Abstract: Aiming at the problem that a single positioning system cannot meet the indoor and outdoor high-precision positioning, GNSS/UWB integrated navigation system is proposed. Firstly, this paper introduces the development and research status of the global satellite navigation system; Secondly, The UWB positioning technology in indoor positioning technology is introduced; Finally, the development and research status of GNSS/UWB integrated navigation are introduced. The advantages and disadvantages of the single outdoor or indoor positioning system in the indoor and outdoor integrated navigation system are compared and analyzed. The results show that a single outdoor or indoor positioning system cannot meet the indoor and outdoor high-precision positioning, an effective method of GNSS/UWB fusion positioning is to improve the positioning accuracy and robustness, which can enhance the real-time in time, continuity in space and reliability in accuracy.

1. Introduction

The global navigation satellite system (GNSS) has the advantages of all-weather positioning, 95% high global coverage, mobile positioning and positioning accuracy not affected by time, high long-term positioning accuracy and error not diverging with time. People pay more and more attention to its role in modern production and life [1-3]. In the open environment, real-time dynamic differential positioning technology (RTK) based on the carrier phase can realize centimeter-level positioning. However, in the complex indoor environment, GNSS will lose some or all of its positioning function when the satellite signal is blocked seriously, and the signal is easily interfered by multi-path environment, so it is difficult to realize real dead angle positioning. As the preferred indoor navigation and positioning technology, UWB (ultra wide band, Chinese Name: ultra wide-band) has a huge bandwidth, with frequencies ranging from 3.1ghz to 0.6GHz. It is precisely because of the large bandwidth that it has the strong anti-multi-path ability and strong penetration performance, at the same time, UWB also has the characteristics of low system complexity, low power spectral density of transmitted signal, insensitive to channel fading, low interception ability and high positioning accuracy. Therefore, the ranging accuracy is better, and the theoretical accuracy can reach the centimeter level, which can meet most positioning accuracy requirements. Therefore, when the GNSS signal is interfered with by blocking or other interference, the UWB positioning technology can be selected for compensation, therefore, we propose to integrate the GNSS navigation system with the UWB indoor positioning technology, use the UWB technology that relies on its nanosecond pulse for signal transmission, compensate for the shortcomings of GNSS in indoor positioning, integrate the local coordinate system of UWB with the global reference frame of GNSS, integrate the excellent indoor positioning performance of UWB with the mature outdoor positioning technology of GNSS, and enhance the real-time, spatial continuity and accuracy reliability of GNSS, thereby realizing indoor outdoor high-precision positioning.

2. Research status at home and abroad

2.1 Development of global satellite navigation system

GNSS is a global space positioning satellite system. As of 2019, the United States' Global Positioning System (GPS), China's BDS Satellite Navigation System (BDS) and Russia's GLONASS have achieved global coverage. The European Union's Galileo positioning system (Galileo) is in the early stages of deployment and is expected to be fully operational by the end of 2022. Global satellite navigation systems not only play an important role in the military but also provide users with real-time location services. Compared with other navigation and positioning systems, global satellite navigation system outdoor positioning has great application advantages. Some countries are also developing regional navigation systems, including India's Regional Navigation Satellite System (IRNSS) and Japan's Quasi-Zenith Satellite System (QZSS) [4].

2.1.1 Status of GPS development

The current GPS satellite constellation consists of 21 operational satellites and 3 spare satellites. As of August 20, 2022, there are a total of 31 GPS satellites in orbit (excluding spare satellites), there are a total of 29 GPS satellites in orbit (excluding spare satellites), which are distributed in 6 orbital planes, and each orbital plane has 4-6 satellites. The GPS satellite constellation distribution is shown in Figure 1.

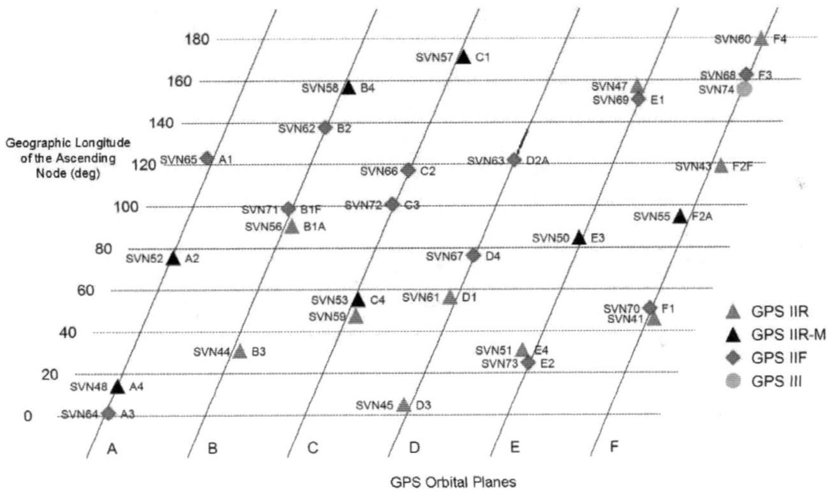

Fig.1 GPS satellite constellation distribution

2.1.2 Development status of BDS

My country's BDS Navigation Satellite System is divided into a three-step development plan: In 2004, BDS-1 completed the establishment of regional active positioning; In 2012, BDS-2 completed its coverage of most of the Asia-Pacific region and officially provided satellite navigation services; In 2020, BDS-3 achieved global passive positioning.

The first generation of the BDS system is the BDS satellite navigation test system, also known as BDS-1, which provides regional positioning services by three satellites. Since 2000, the system has mainly provided navigation services within China. The second generation of the BDS system is the BDS satellite navigation system, also known as BDS-2. As of November 2011, BDS II contains 10 satellites and has begun to enter service in China. In November 2012, the second-generation BDS system began to provide users with regional positioning services in the Asia-Pacific region. The BDS

III navigation system is a hybrid constellation system consisting of "3GEO+3IGSO+24MEO" based on the BDS-2 regional satellites [8,9]. As of December 16, 2019, 28 third-generation in-orbit navigation satellites have been launched, marking the completion of the deployment of the core constellation of the BDS-3 global system, as shown in Table 2. The system covers countries along the "Belt and Road" in 2018, and the construction is completed in 2020 to provide global positioning services, and a comprehensive Positioning, Navigation, and Timing system (PNT) with BDS is to be built as the core in 2035.

Table 1 List of BDS-3 satellites

Serial number	Satellites	Launch date	Satellite orbits
1	BDS-3-M1	2017/11/5	MEO
2	BDS-3-M2		MEO
3	BDS-3-M3	2018/1/12	MEO
4	BDS-3-M4		MEO
5	BDS-3-M5	2018/2/12	MEO
6	BDS-3-M6		MEO
7	BDS-3-M7	2018/3/30	MEO
8	BDS-3-M8		MEO
9	BDS-3-M9	2018/7/29	MEO
10	BDS-3-M10		MEO
11	BDS-3-M11	2018/8/25	MEO
12	BDS-3-M12		MEO
13	BDS-3-M13	2018/9/19	MEO
14	BDS-3-M14		MEO
15	BDS-3-M15	2018/10/15	MEO
16	BDS-3-M16		MEO
17	BDS-3-G1	2018/11/1	GEO
18	BDS-3-M17	2018/11/19	MEO
19	BDS-3-M18		MEO
20	BDS-3-IGSO1	2019/4/20	IGSO
21	BDS-3-IGSO2	2019/6/25	IGSO
22	BDS-3-M23	2019/9/23	MEO
23	BDS-3-M24		MEO
24	BDS-3-IGSO3	2019/11/5	IGSO
25	BDS-3-M21	2019/11/23	MEO
26	BDS-3-M22		MEO
27	BDS-3-M19	2019/12/16	MEO
28	BDS-3-M20		MEO
29	BDS-3-G2	2020/3/9	GEO
30	BDS-3-G3	2020/6/23	GEO

2.1.3 Landslide monitoring experiment conclusion

As shown in Figure 2, it shows the accuracy distribution map of the BeiDou system satellite, and with the increase in the number of observation calendars, the robustness of observation is enhanced and the accuracy is continuously improved.

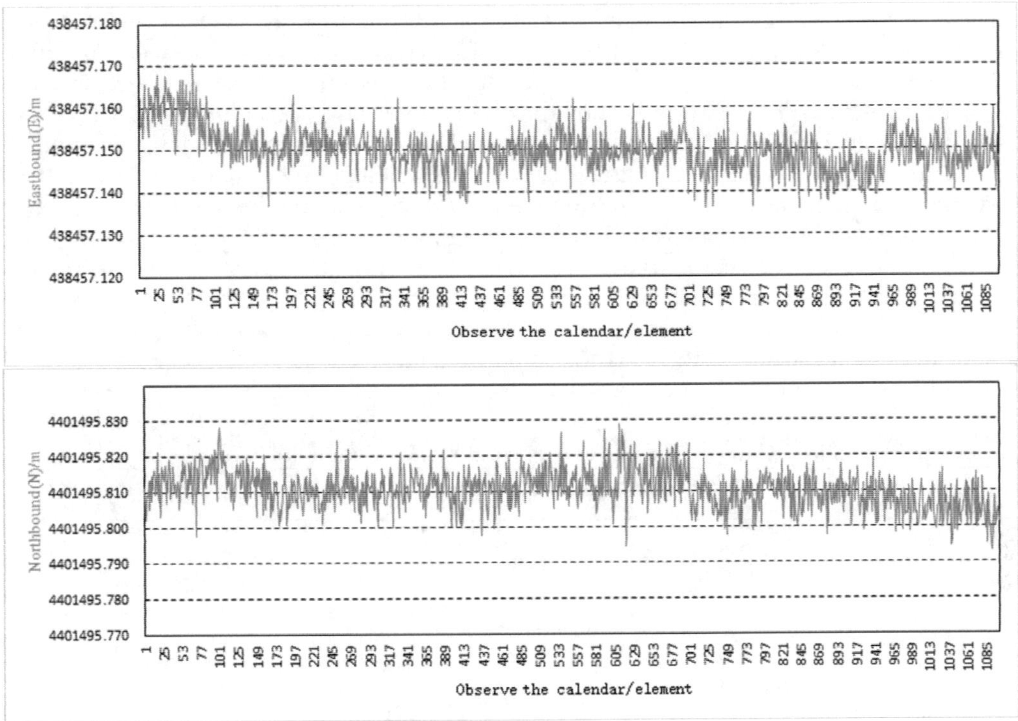

Fig.2 GPS time series

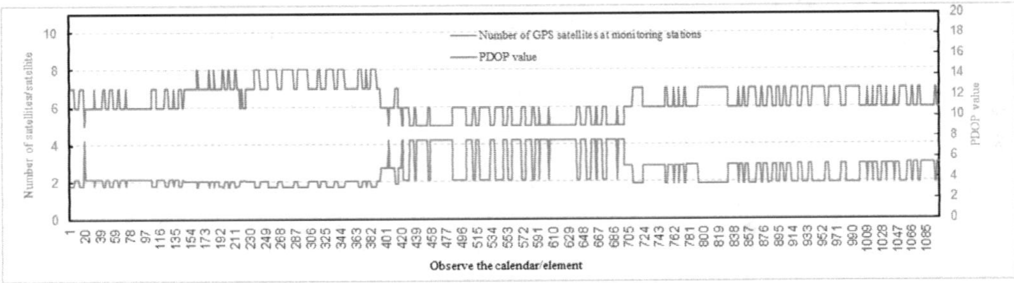

Fig.3 Spatial geometric intensity factors of BeiDou satellite distribution

As shown in the Figure, yellow indicates the spatial geometric intensity factor (PDOP) of the satellite distribution of the BeiDou system, and the better the general satellite distribution, the smaller the PDOP value, and the ideal state is generally less than 3. It can be seen that with the increase of the observation calendar, the PDOP value tends to 3. In addition, the blue image indicates the status of the number of visible satellites of the BeiDou system under a certain height angle, and with the increase of the observation calendar, the number of satellites tends to stabilize at about 6 to 7.

2.2 The development of ultra-broadband indoor positioning technology

The Indoor Positioning System (IPS) function is to display the location information of people or objects in an indoor environment, and its purpose is to grasp the position of the target object in real time.

In recent years, indoor and outdoor seamless positioning technology has become a research focus. The outdoor positioning method based on GNSS (Global Navigation Satellite System) can achieve sub-meter positioning, based on WLAN (wireless local area network)[5], WIFI (wireless fidelity)[6], Bluetooth[7], ultra-broadband indoor positioning method[8] and other sensors can also obtain sub-meter positioning accuracy. However, a high-precision, fast and seamless positioning method is even more difficult to achieve. Commonly used indoor positioning technologies mainly include the following: Reference [9] introduced a WiFi and Beacon fusion indoor positioning method, and described its excellent advantages and broad application prospects in detail; Yang Di et al. introduced an ultra-broadband indoor positioning technology, expounded the positioning principle and method, and pointed out the current problems of ultra-broadband positioning[10]; Reference [11] introduced the RFID indoor positioning technology, improved the positioning accuracy by improving the algorithm, mainly introduced several algorithms and improved algorithms, and verified their accuracy through experiments; Reference [12] introduced the infrared and ultrasonic indoor positioning technology, designed the ultrasonic and infrared system positioning scheme, analyzed the error source in detail, and realized the positioning and tracking of the moving target. The comparison of various wireless positioning technology parameters is shown in Table 2, and the performance comparison of the positioning system is shown in Table 3.

Table 2 Comparison of various wireless positioning technologies

Scheme	Type/Location Mechanism	Precision	Advantage	Disadvantage	cost
GNSS	Geometric positioning	Meter level	Easy to use outdoors; High precision	Signal is easy to block; It is not available indoors	High
Infrared	Geometric positioning	Centimeter-level	High precision	Poor anti-interference ability; Requirements are visible	High
Ultrasound	Geometric positioning/Echo measurement	Centimeter-level	High precision	Fast decay and small range	High
ZigBee	Geometric positioning/Based on RSSI	Meter level/3~5m	Low power consumption	Poor anti-interference ability; Poor stability	Low
TC-OFDM	Geometric positioning	Meter level	The accuracy is high in the meter level	New communication system	High
UWB	Geometric positioning/TDOA	Centimeter-level/0.15m	High precision; Anti-interference	Small coverage	High
Geomagnetic	Scene Recognition	Meter level	High precision	Prior sampling is required; Poor stability	Low
WIFI	Geometric positioning/Scene Recognition	Meter level~Ten meters class	Easily accessible; Strong communication ability	Unstable accuracy; Energy consumption	Low
Bluetooth	Geometric positioning/Scene Recognition/Proximity Perception	Meter level	Low power consumption; High security	Poor anti-interference ability; The communication range is small	Low
Cellular Network	Geometric positioning/Scene Recognition/Proximity Perception	Ten meters class	Easily accessible	Dependence on base station density; Scene Recognition requires sampling	Low
Sensor Inertial Navigation	Proximity Perception	Meter level	Strong anti-interference ability; Low deployment costs	Error accumulation	Extremely low
Radio Frequency Identification	Proximity Perception	Centimeter-level	High precision	The range is short and the tag has no communication ability	Low
Computer Vision	Proactive targeting	Subminomic	High precision/	A large amount of	High

		Independence	operation; Poor stability; Required line of sight;

2.3 UWB simulation testing

To evaluate the accuracy of UWB positioning, the simulation test results are shown in the figure compared to the distance error in the x and y directions, and the random ranging value under the constraint of the same ranging variance is constructed.

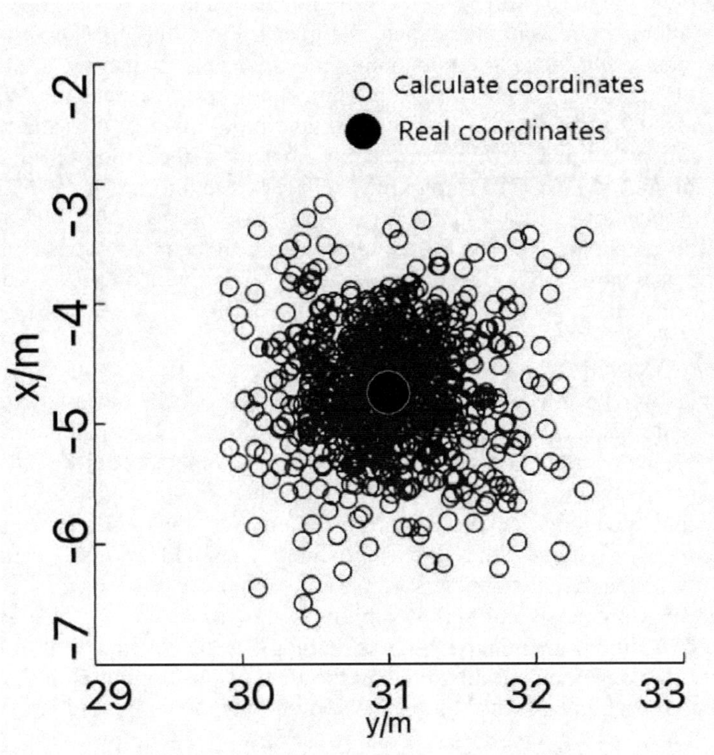

Fig.4 Simulation test of random-ranging values

Table 3 Performance comparison of various indoor positioning systems

indoor positioning systems	Precision	Penetration	Anti-interference ability	Layout Complexity	Cost
Infrared	4	0	0	5	2
Ultrasound	5	1	3	2	5
RFID	3	3	2	3	5
Bluetooth	3	3	2	3	1
WiFi	1	3	5	1	1
ZigBee	2	4	3	2	3
UWB	5	5	4	3	4
Geomagnetic Field	4	4	3	3	4

At present, with the development of the social economy, people's demand for location services has become stronger and stronger, which has extended from outdoor to indoor environments. UWB is an emerging communication technology that uses pulses with a width of nanoseconds as wireless communication signals, especially in terms of time resolution and positioning accuracy can be accurate to the centimeter level. In addition, it also has good performance in anti-multi-path ability and penetration, suitable for high-precision indoor positioning services.

UWB technology has the characteristics of high data transmission rate, strong anti-multi-path interference ability, low power consumption, low cost, strong penetration ability, low interception rate and shared spectrum with other existing wireless communication systems[13]. UWB technology started from the pulse communication technology that emerged in the 1960s. It uses ultra-wide base band pulses with an extremely wide spectrum for communication and is mainly used in military radar, positioning and communication systems with low interception rate/low detection rate. In 2001, China released the national "863 plan" for the first time to take ultra-broadband as the research content of wireless communication common technology and innovative technology. In 2002, the Federal Communications Commission (FCC) approved the wireless technology for civilian communications. Subsequently, Japan opened the ultra-wideband frequency band in 2006. In 2007, the IEEE officially released the 8025.15.4a standard, and the pulsed ultra-wideband technology was selected as one of the main physical layer standards. ISO officially passed the first one. In 2019, Apple's iPhone 11 is equipped with a U1 chip that supports UWB technology.

2.4 Development of GNSS/UWB Integrated Navigation System

The integrated navigation system refers to the combination of two or more navigation methods, the fusion of their navigation data, and the use of certain algorithms such as Kalman filtering to obtain a fusion result. Usually, the combination between inertial navigation and satellite navigation is the most widely used.

Because the global navigation and positioning system and the inertial navigation system can achieve complementary advantages, they have been widely studied by many scholars at home and abroad. Compared with single navigation systems in complex environments, GNSS/UWB fusion provides users with high-precision and high-reliability navigation information [14-16]. Regardless of the development trend of positioning and navigation technology or the development of market application requirements, one of the development directions in the field of positioning is to develop a positioning and navigation system that integrates multiple positioning technologies and can be used in indoor, outdoor, and indoor and foreign borders and other scenarios. To comply with the macro trend of integrated development of positioning and navigation, we have developed high-precision positioning system products integrating satellite navigation and positioning, UWB high-precision indoor positioning, inertial navigation, and other technologies, and indoor and outdoor integrated positioning module products with high precision, low power consumption and small size, to truly realize the seamless integration of indoor and outdoor positioning and navigation from the upper layer to the lower layer.

GNSS/UWB integrated navigation technology uses distributed Kalman filtering for data fusion based on UWB signal arrival time difference, signal incidence angle and deployment topology to perform high-precision position calculation. At the same time, it optimizes the calculation results based on map matching, path optimization algorithm and artificial intelligence decision model. In the application layer, map matching, trajectory fitting and seamless connection of indoor and outdoor maps are realized, and automatic switching of indoor and outdoor maps and coordinate systems is realized. The combination mode is simple and has certain redundancy. The UWB base station equipment can be arranged step by step from outside to inside without prior arrangement. The coordinates of the base station can be iterated to the indoor through the outdoor positioning data without prior measurement. Based on tag equipment and cloud platform, distance information can be transmitted to the platform in real time through 4G network to locate personnel and realize online monitoring of personnel. This method has the characteristics of fast layout, strong flexibility and high positioning accuracy.

From the perspective of GNSS positioning technology, GNSS/UWB integrated navigation mode can be divided into: standard point positioning (SPP)/UWB integrated technology based on pseudo range, precise point positioning (PPP)/UWB integrated technology based on precise point positioning (PPP)/UWB integrated technology, Differential global positioning system (DGPS)/UWB integrated technology and real time dynamic carrier phase difference (RTK)/UWB integrated navigation technology.

3. Research status of GNSS/UWB integrated navigation

At present, there are three research directions for seamless positioning. One is to combine GNSS technology with indoor positioning technology to achieve seamless positioning. At present, a large number of researches are mainly focused on this aspect. Other is to solve the problem of seamless positioning by introducing GNSS signals indoors. The representative technology is Japan's indoor messaging system (IMES). The principle of this system is similar to that of a pseudo satellite, and the positioning signal with the same structure as the GNSS signal is transmitted to realize the seamless connection between indoor and outdoor positioning. The other one is that through the realization of a regional positioning technology, the indoor and outdoor positioning information will be unified under a technical system.

Scholars at home and abroad have made a lot of research on the GNSS/UWB combination. Opshaug loosely coupled GNSS/UWB combined data to obtain GNSS Compared with GNSS single system, UWB integrated navigation system has a higher spatial distribution of satellites, an increase of redundant observation and positioning success rate [17]; Tanigawa et al. directly fused the localization results of GNSS and UWB [18]; Wu Tanshi proposed to use Helmert variance component estimation to determine the reasonable weight of the combined system view measurement [15]; Hu Quan et al. used the trajectory correction algorithm to perform combined positioning on the plane coordinates of UWB, real-time speed and azimuth information of GNSS system [19].

Through the above research of domestic and foreign scholars and the analysis of the research status of seamless positioning, the integrated navigation system is still in the development stage, and there is still great research value in the field of indoor and outdoor integrated seamless positioning in the future.

4. Conclusions and future work

This paper briefly introduces the current global satellite navigation system and UWB indoor positioning technology, and introduces the development and research status of GNSS/UWB integrated navigation. Through comparative analysis, the following conclusions can be obtained: (1) UWB indoor positioning technology can compensate for the problem of insufficient positioning accuracy of GNSS indoor; (2) GNSS/UWB integrated navigation system can effectively improve positioning accuracy and positioning robustness; (3) Indoor and outdoor seamless positioning integration has great potential.

With the improvement of positioning accuracy, a single positioning system can not meet the seamless switching of multiple scenes. We propose to integrate GNSS/UWB technology, i.e. integrate the local coordinate system of UWB with the global reference frame of GNSS, integrate the excellent indoor positioning performance of UWB with the mature outdoor positioning technology of GNSS, and enhance the real-time, spatial continuity and accuracy reliability in time.

References

[1] LEICK A, RAPOPORT L, TATARNIKOV D. (2015) *GPS satellite surveying* [M]. John Wiley & Sons.

[2] Wang N, Song W, Guo H. (2019) Research on GNSS/UWB high-precision indoor and outdoor combined positioning method [J]. *Communication power supply technology*, 36(10): 22-23.

[3] Li Z (2013). GPS measurement and data processing.

[4] Wang J, Zhang A. (2017) Principles and applications of satellite positioning. *Beijing: Surveying and Mapping Press.*
[5] Khalajmehrabadi A, Gatsis N, Akopian D. (2017) Modern WLAN fingerprinting indoor positioning methods and deployment challenges. *IEEE Commun. Surv.* Tutor.19, 1974–2002.
[6] Wang J, Hu A, Liu C, Li X. (2015) A floor-map-aided WiFi/ pseudo- odometry integration algorithm for an indoor positioning system. Sensors 15, 7096–7124.
[7] Chen R, Chen L. (2017) The current status and challenges of the development of indoor positioning technology based on smartphones [J]. *Acta Geodaetica et Cartographica Sinica.* 15, 1316-1326.
[8] Alsindi N, Alavi B, Pahlavan K. Spatial Characteristics of UWB TOA -based Ranging in Indoor Multi-path Environments. In Proceedings of the IEEE 18th International Symposium on Personal, *Indoor and Mobile Radio Communications, Athens*, Greece, 3–7 September 2007.
[9] Peng Y (2018) Wireless WiFi is integrated with Bluetooth Beacon positioning [J].
[10] *Telecom World* (02): 4-5.
[11] Yang D, Tang X, Li B, et al. (2015) A Review on Indoor Positioning Technology Based on Ultra-Wideband [J]. *GNSS World of China*, 40(5): 34-40.
[12] Xu L. (2013) Research and design of indoor positioning system based on RFID technology [D]. *Nanjing University of Science and Technology.*
[13] Zeng X. (2017) Research and implementation of indoor real-time positioning system based on ultrasonic and infrared [D]. *Huaqiao University.*
[14] Li X, Yang H. (2016) Introduction to IoT Education Engineering [M]. *Wuhan: Huazhong University of Science and Technology Press.*
[15] An Emergency Seamless Positioning Technique Based on *ad UWB Networking Using Robust EKF.*
[16] Wu T. (2017) Research on key technologies of GPS/UWB combined positioning [J]. *Geomatics & Spatial Information Technology,* 40(05): 137-139+142.
[17] Liu Z. Research on indoor and outdoor seamless positioning technology based on UWB_GNSS_MIMU
[18] Opshaug, G.R., Enge P. Integrated GPS and UWB navigation system: (motivates the necessity of non-interference). *Ultra Wideband Systems and Technologies*, 2002. Digest of Papers. 2002 IEEE Conference on, 2002
[19] Glenn D. Macgougan Real-time kinematic surveying using tightly-coupled GPS and ultra-wideband ranging. [D]. *University of Calgary*, 2009
[20] Hu Q, He T, Jia T, Li Y, Tang Y. (2018) A UWB/GPS combination positioning method [J]. Journal of Nanjing University of Science and Technology, 42(01): 76-81.

Accuracy Analysis of Timing Comparison for BDS-3 and GPS Based on Satellite Common View Algorithm

Mingduan Zhou, Xu Ji, Muyuan Zhang, Ye Tian, Leyi Shang

School of Geomatics and Urban Spatial Informatics, Beijing University of Civil Engineering and Architecture, Beijing 102616, China

E-mail: mdzhou@bucea.edu.cn

Abstract: With the networking completion of the Beidou navigation satellite system (BDS-3), the BDS-3 system was opened in the whole world on July 31, 2020. It is important practical significance to analyze the performance of the BDS-3 system and the GPS in terms of satellite common view timing comparison. In order to analyze the accuracy of timing comparison for BDS-3/GPS global navigation satellite system common view time, the comparison experimental of BDS-3/GPS with medium and long baseline is carried out based on satellite common view algorithm by using the National Metrology benchmark system of atomic time scale maintained by National Institute of Metrology (NIM), China. The timing model and algorithm of the BDS-3/GPS satellite common view algorithm are deduced in detail, and the stability of receiver clock difference and accuracy analysis of timing comparison for the BDS-3/GPS based on the satellite common view algorithm are compared and analyzed. The experimental results show that the stability of receiver clock difference of BDS-3 and GPS is respectively ns-level, but the stability of receiver clock difference of BDS-3 system is slightly better than that of GPS; the accuracy analysis of timing comparison for BDS-3 and GPS based on satellite common view algorithm is respectively better than 24 ns, but the accuracy analysis of timing comparison for BDS-3 system is slightly better than that of the GPS.

1. Introduction

With the rapid development of the modernization of global navigation satellite system (GNSS), the application demand for smart grid system, communication system, national defense system and other industries for GNSS timing technology are increasing. In 1980, the GPS satellite timing technology based on the common view method (CV) was firstly proposed by the US national bureau of standards using a single channel C/A code common view receiver for the time comparison between atomic clocks in the international punctuality laboratory [1]. Since GPS C/A code was proposed for timing comparison and transmission, GNSS satellite common view time technology has developed rapidly. In recent years, GNSS satellite common view timing technology has become one of the focus topics in some industrial applications. Virtual common view reference station technology was proposed to solve the problem of satellite common view timing method with users far away from the reference station timing accuracy degradation [2]. The GPS common view timing accuracy of different baseline lengths by using the satellite common view timing algorithm was analyzed [3]. Considering the characteristics of various types of the constellation for the BDS-3 system, the single satellite timing accuracy was analyzed based on different satellites' orbits consisting of GEO/IGSO/MEO [4]. The power grid time synchronization method for the remote calibration of power grid time and frequency was proposed by using satellite common view technology [5]. An improved satellite common view method was proposed and a set of high-precision remote timing common view comparison system was built based on the BDS/GPS under the limitations of the standard satellite common view method [6]. The standard time remote high-precision method based on the principle of satellite common view was discussed and the corresponding system and terminal were developed [7]. The satellite common view timing method device was developed, and the space-based time synchronization network experimental system was built in the implementation method of the space-based time synchronization network [8]. Feasibility of implementing satellite common viewing technology using single-frequency receivers instead of dual-frequency receivers for high-precision time transfer applications in telecommunication networks [9]. The experimental results of Asia Europe common view timing comparison of BDS-3 satellite with a baseline length of about 7500km were analyzed [10]. To realize the real-time traceability and real-time synchronization of the standard to the benchmark, the satellite common view data with UTC and NIMDO were compared [11]. Based on the satellite common view comparison technology and the clock control method, the local clock was used to reproduce a high-precision time-frequency signal synchronized with UTC (NTSC). When the crystal oscillator is used as the local clock, the reproduced signal is compared with UTC (NTSC) [12]. The UTC (NTSC) remote high-precision reproduction system was developed to solve the discontinuity in the measurement of the standard satellite common view method [13]. With the networking completion of the Beidou navigation satellite system (BDS-3), BDS-3 was opened in the whole world on July 31, 2020. It has important practical significance in national defense and military fields because of its autonomy, controllability and security. However, the performance analysis of timing comparison between BDS-3 system and GPS based on satellite common view timing algorithm is slightly reported recently. Considering the low-cost, high-precision and convenient operation of the GNSS satellite common view timing device, it is of great significance to solve the demand for high-precision time synchronization in a smart grid system, communication system, national defense system and other industrial applications. Therefore, the timing comparison experimental of BDS-3/GPS with about 27 km was carried out based on a satellite common view algorithm by using the National primary standard of measurement of atomic time scale maintained by the National Institute of Metrology (NIM), China. The experiment of timing comparison for the BDS-3/GPS was compared and analyzed. The result for the stability of receiver clock difference and accuracy analysis of timing comparison for BDS-3 and GPS based on satellite common view algorithm is of great significance.

The structure of this paper is as follows. Firstly, the timing model and algorithm for the satellite common view method are given, and then an experimental observation scheme is designed for satellite common viewing data acquisition and timing comparison based on GPS and BDS systems, and the two sets of experimental results are analyzed in depth. Finally, the corresponding conclusions are summarized.

2. Timing model and algorithm of satellite common view method

Satellite common view method is a timing comparison algorithm between two-time service stations based on a global navigation satellite system (GNSS). The two-timing stations for timing comparison strictly observe the satellite common view timetable and establish a data transmission network to realize common view data exchange. The international Bureau of Weights and Measures (BIPM) updates the satellite common view timetable every six months and release it to the public. The standard satellite common view time interval is 16 minutes. Among them, 13 minutes are used for satellite data acquisition, 2 minutes for data processing, and 1 minute for the arrival of the next common view time. The satellite common view time method technology is that two or more observation stations located in different locations observe the same navigation satellite at the same time, calculate the time difference between the navigation satellite and the observation station receiver at the corresponding observation time, and obtain the time difference between the two observation stations through real-time or post common view data exchange. The working principle for satellite common view time method is shown in Figure 1.

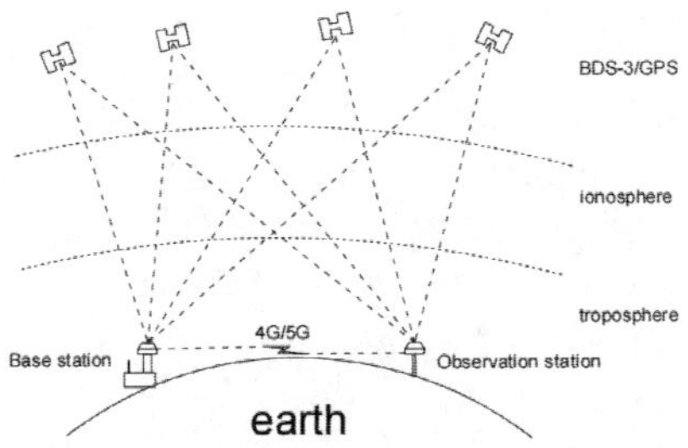

Figure 1. Working principle of satellite common view time method

According to Figure 1, it is assuming that the length of the baseline formed between the reference station (represented by subscript *a*) and the observation station (represented by subscript *b*) is within 100 km, and for a satellite common view observation time, the reference station and the observation station synchronously observe the satellite, the pseudo-range observation Equation is listed as follows:

$$\rho_a^j = R_a^j - c\cdot\Delta t + c\cdot\Delta t_a^j + Ion_a^j + Trop_a^j + \varepsilon_a^j \tag{1}$$

$$\rho_b^j = R_b^j - c\cdot\Delta t + c\cdot\Delta t_b^j + Ion_b^j + Trop_b^j + \varepsilon_b^j \tag{2}$$

Where ρ_a^j, ρ_b^j, ρ_a^j and ρ_b^j are the pseudo-range observations from the satellite j to the reference station and the observation station respectively; c is the speed of light; Δt is the satellite clock difference of the satellite j; Δt_a^j and Δt_b^j are receiver clock difference of reference station and observation station respectively; Ion_a^j and Ion_b^j are ionospheric delay error of reference station and observation station respectively; $Trop_a^j$ and $Trop_b^j$ are tropospheric delay error of reference station and observation station; ε_a^j and ε_b^j are the measurement noise of reference stations and observation station respectively; R_a^j and R_b^j are the geometric distance of the satellite j to the reference station and the observation station respectively. The calculation formula of R_a^j and R_b^j are listed as follows:

$$R_a^j = \sqrt{(X^j - X_a^j)^2 + (Y^j - Y_a^j)^2 + (Z^j - Z_a^j)^2} \tag{3}$$

$$R_b^j = \sqrt{(X^j - X_b^j)^2 + (Y^j - Y_b^j)^2 + (Z^j - Z_b^j)^2} \tag{4}$$

Where $(X_a^j,\ Y_a^j,\ Z_a^j)$ is the three-dimensional spatial coordinates of the reference station; $(X_b^j,\ Y_b^j,\ Z_b^j)$ is the three-dimensional spatial coordinates of the observation station; $(X^j,\ Y^j,\ Z^j)$ is the three-dimensional spatial coordinates of the satellite j at common view time.

By subtracting Equation (1) from Equation (2), the satellite clock difference of the navigation satellites is eliminated and the result of the single-satellite common view between the reference station and the observation station is obtained:

$$\Delta T_{ab}^j = \Delta t_a^j - \Delta t_a^j = \frac{1}{c}\cdot(\Delta\rho_{ab}^j - \Delta R_{ab}^j - \Delta Ion_{ab}^j - \Delta Trop_{ab}^j - \Delta\varepsilon_{ab}^j) \tag{5}$$

Where $\Delta\rho_{ab}^j = \rho_a^j - \rho_b^j$; $\Delta R_{ab}^j = R_a^j - R_b^j$; $\Delta Ion_{ab}^j = Ion_a^j - Ion_b^j$; $\Delta Trop_{ab}^j = Trop_a^j - Trop_b^j$;

$\Delta\varepsilon_{ab}^{j}=\varepsilon_{a}^{j}-\varepsilon_{b}^{j}$.

In order to improve the accuracy and reliability of the satellite common view timing comparison results. If a navigation satellite is observed synchronously at a certain satellite common-viewing observation time, and the multi-satellite common view results are obtained according to the weighted fusion process according to the satellite height angle of the satellite j, where $j=1,2,\cdots,m$, and the single-satellite common view result is listed as follows:

$$\Delta T_{ab}=\sum_{j=1}^{m}\left[P^{j}\cdot\Delta T_{ab}^{j}\right]\Big/\sum_{j=1}^{m}P^{j} \tag{6}$$

Where, the weight matrix is based on the satellite altitude angle weighting strategy [12-13]. Among them, a simple method of determining the weighting for satellites is listed as follows:

$$P^{j}=\begin{cases}2\times\sin(e^{j})\ , & 0°<e^{j}\leq 30°\\ 1\ , & 30°<e^{j}\leq 90°\end{cases} \tag{7}$$

Where, e^{j} is the satellite altitude angle of the navigation satellite j.

3. Experiment and analysis

In order to verify and analyze the accuracy of BDS-3/GPS navigation satellite common view timing comparison, a timing receiver device is set up on the control point of an experimental field as an experimental observation station (the sub-station), and based on the atomic time-standard National metrology reference system maintained by NIM was adopted as the reference station for the timing comparison experiment (the main station, IP: 47.104.8.207, port: 8119). The data sampling rate of the timing receiver device of the experimental observation is set to 1 Hz, the cut off altitude angle of the satellite is set to 15°. The baseline formed between the experimental observation station and the reference station is about 27 km. The satellite common view data was based on the common GNSS generic time transfer standard (CGGTTS) issued by the Consultative Committee for Time and Frequency (CCTF), and the use of mobile communications 4G/5G networks to achieve data exchange.

In order to evaluate and analyze the accuracy of timing comparison for BDS-3/GPS based on satellite common view method, two experimental observation schemes are designed as follows: (1) Scheme 1: Design a satellite common-view data acquisition and timing comparison experimental observation scheme based on the BDS-3 system. (2) Scheme 2: Design a satellite common-view data acquisition and timing comparison experimental observation scheme based on the GPS.

Firstly, in order to evaluate and compare the difference in clock difference between the timing receiver based on the BDS-3 system and the GPS, the timing receiver clock difference data between the second pulse time is used to measure the local clock output and the GNSS time of the navigation satellite signal reaching the receiver output (monitored by the time interval counter module), selected continuous one-hour monitoring results for numerical statistics and analysis, to obtain the timing receiver clock difference monitoring results in Figure 2. Among them, Figure 2(a) shows the receiver clock difference time series analysis results for one consecutive hour (12:00:00-13:00:00 on August 11, 2021) for the BDS-3 system, and Figure 2(b) gives the receiver clock difference time series analysis results for one consecutive hour (12:00:00-13:00:00 on August 27, 2021) for the GPS.

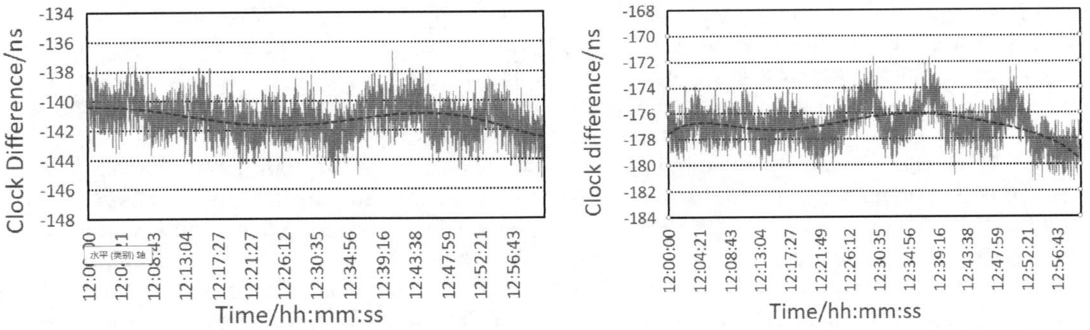

(a)Analysis results of receiver clock difference time series based on BDS-3 system

(b)Analysis results of receiver clock difference time series based on the GPS

Figure 2. Clock difference monitoring results of the timing receiver

As can be seen from Figure 2, the maximum value of the clock difference of the timing receiver based on the BDS-3 system is -133.4 ns, the minimum value is -147.5 ns, and the average value is -141.7 ns; the maximum value of the clock difference of the timing type receiver based on the GPS is -170.0 ns, the minimum value is -190.6 ns, and the average value is -181.7 ns. This shows that for the receiver clock difference monitoring results monitored by the time interval counter, the average clock difference of the timing receiver based on the BDS-3 system (-141.7 ns) is better than the GPS (-181.7 ns).

In order to further evaluate and compare the stability of the timing receiver clock difference based on the BDS-3 system and the GPS, for the receiver clock difference data monitored by the time interval counter, the monitoring results of the whole day (the BDS-3 system selects 00:00:00-24:00:00 on August 12, 2021 and the GPS selects 00:00:00-24:00:00 on August 28, 2021) for numerical statistics and analysis, and the single-day monitoring results of the timing receiver clock difference are obtained. And the selected receiver clock difference single-day monitoring results from the internal compliance accuracy (Standard deviation, STD) of the receiver clock difference stability assessment and analysis of different satellite systems: the numerical statistical analysis of the single-day monitoring results of the timing receiver clock difference is shown in Table 1. The calculation formula of STD is listed as follows:

$$STD = \sqrt{\frac{\sum_{i=1}^{n}(X_i - \bar{X})^2}{n-1}} \tag{8}$$

Where, X_i is the monitoring amount, $\bar{X}$ is the average of the monitoring quantity, and n is the number of monitoring epoch.

Table 1. Numerical statistical analysis of single-day monitoring results of receiver clock difference

Indicator points	BDS-3	GPS
maximum/ns	-133.5	-171.9
minimum/ns	-147.1	-191.5
average /ns	-141.7	-180.9
STD/ns	1.8	2.1

As can be seen from Table 1, the maximum value of the single-day monitoring result of the timing receiver clock difference for the BDS-3 system is -133.5 ns, the minimum value is -147.1 ns, the average value is -141.7 ns, and STD value is 1.8 ns; the maximum value of the single-week monitoring result of the timing receiver clock difference for the GPS is -171.9 ns, the minimum value is -191.5 ns, and the average value is - 180.9ns, STD value is 2.1 ns. The single-day monitoring results of the receiver clock difference monitored by the time interval counter show that the receiver clock difference stability of both the BDS-3 system and the GPS are in the ns-level, but the timing receiver clock difference stability of the BDS-3 system is slightly better than the GPS.

Besides, in order to evaluate and compare the accuracy of the satellite common view method based on the BDS-3 system and the GPS, the satellite common view time of the BDS-3 system is selected as 00:02:00-23:42:00 on August 12 (there is a common viewpoint every 16 minutes, a total of 90 common viewpoints) and the GPS satellite common view time is August 28, 00:02:00-23:42:00 (every 16 minutes there is a common view point, a total of 90 common viewpoints) The single-day monitoring data were analyzed with satellite common view method of timing comparison results (Figure 3).

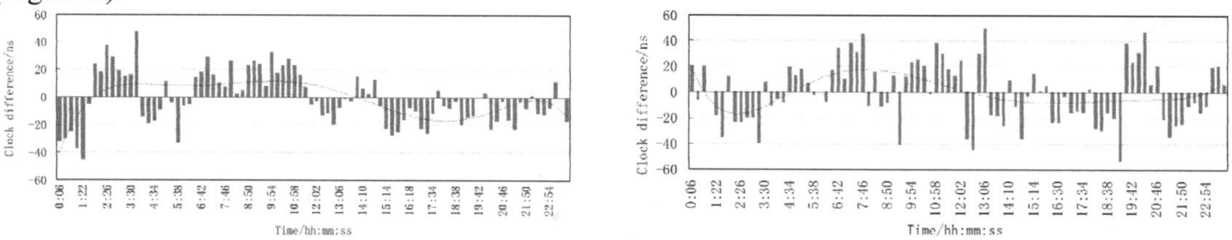

(a) Comparison results of the BDS-3 system-based satellite common view method

(b) Comparison results of the satellite common view method based on GPS

Figure 3. single-day monitoring data were analyzed with satellite common view method of timing comparison results

In order to comprehensively evaluate and compare the accuracy of the satellite common view method timing comparison results based on the BDS-3 system and the GPS, the numerical statistics and analysis of the selected single-day satellite common view method of timing comparison results based on the BDS-3 system and the GPS are shown in Table 2.

Table 2. Numerical statistical analysis of timing comparison results by satellite common view method

Indicator points	BDS-3	GPS
Maximum/ns	47.0	50.0
Minimum/ns	-45.0	-53.0
Average/ns	-1.0	-0.1
STD/ns	18.6	23.2

As can be seen from Table 2, the maximum value of the satellite common view method timing comparison result based on the BDS-3 system is 47.0 ns, the minimum value is -45.0 ns, the average value is -1.0 ns, STD is 18.6 ns, the maximum value of the satellite common view method timing comparison result based on the GPS is 50.0 ns, the minimum value is -53.0 ns, the average value is -0.1 ns, and the STD is 23.2 ns. The results show that the accuracy of the BDS-3 system and the GPS are both better than 24 ns, but the accuracy of the BDS-3 system is slightly better than that of the GPS.

4. Conclusions and future work

In this paper, the timing model and algorithm of BDS-3/GPS satellite common view algorithm are deduced in detail, and the stability of receiver clock difference and accuracy analysis of timing comparison for BDS-3/GPS based on satellite common view algorithm are compared and analyzed. The comparison testing of BDS-3/GPS with the length of baseline about 27 km is carried out based on satellite common view algorithm by using the National Metrology benchmark system of atomic time scale maintained by NIM institution, the conclusions are summarized as follows:(1) The stability of receiver clock difference of BDS-3 and GPS is respectively ns-level, but the stability of receiver clock difference of BDS-3 system is slightly better than that of GPS. (2) The accuracy analysis of timing comparison for BDS-3 and GPS based on satellite common view algorithm is respectively better than 24 ns, but the accuracy analysis of timing comparison for BDS-3 system is slightly better than that of GPS.

It is worth noting that the timing comparison experiment of BDS-3/GPS （based on satellite common view algorithm is only considered under the same satellite common view time, and the experimental results are analyzed quantitatively. Future work is given as follows: (1) The effects of GNSS satellite common view timing comparison experimental results will be deeply compared and analyzed under the different environmental factors such as atmosphere, electromagnetic field, temperature and humidity. (2) BDS-3/GPS navigation satellite common view timing comparison experiment will be carried out under the short baseline, long baseline and ultra-long baseline respectively.

References

[1] D.W. Allan and M.A Weiss 1980 *Symp* Accurate time and frequency transfer during common-view of a GPS satellite. *Proceedings of the 1980 Frequency Control Symposium*, 334-346.10th International Conference on European Frequency and Time
[2] Xu L 2016 Realization of a GNSS one-way timing method. *Journal of Time and Frequency*, 39(4): 290-300.
[3] Wang Y 2017 Algorithm of GPS common-view timing on different baseline lengths. *Journal of Navigation and Positioning*, 5(4), 41-45.
[4] Zhang D and Zheng Z, 2019 Accuracy analysis of single satellite timing of Beidou Navigation Satellite System. *Bulletin of Surveying and Mapping*, 19(4), 7-10.
[5] Zhao S 2016 Research on time measurement and synchronization using satellite common-view in power grid. *Computer Measurement & Control,* 24(12), 49-52.
[6] Chen R 2016 Study on remote time comparison based on improved satellite common-view method. *Chinese Journal of Scientific Instrument,* 37(4), 757-763.
[7] Liu Y 2018 A method for remote reproduction of standard time based on GNSS common-view technology. *Journal of Time and Frequency, vol*41(3), 194-199.
[8] Duan X 2019 Time synchronization network in UPIoT based on satellite common-view technology. *Sichuan Electric Power Technology*, 42(5), 1-4.
[9] Zeng W 2020 Analysis of synchronization performance between two stations with satellite single-frequency close-range common view and dual-frequency one-way timing. *Journal of Time and Frequency*, 43(2), 101-112.
[10] Wang W *2020* Analysis of common view time comparison by BDS-3 satellite. *Journal of Astronautics*, 41(5). 569-577.
[11] Wang L 2017 Research on the remote calibration method of time frequency based on GNSS common-view way. *Shanghai: Shanghai Jiao Tong University.*
[12] Wang, J, Zhang, A (2017). Principles and applications of satellite positioning. *Beijing: Surveying and Mapping Press.*
[13] Zhou M, Zhao J and Zang Z 2021 Single-epoch timing method, device and system based on Beidou satellite signal: CH. Patent 202111051613.4[P].2021-06-17.
[14] Chen P and Zhang Y (2018). GNSS time offset determination. *Beijing: Surveying and Mapping Press.*

Study on Short-term and Imminent Precipitation Prediction Model and Accuracy Evaluation Based on Depth Learning

Yuanyuan Liao[1,a], Gang Yin[2,b]

[1]College of Information Science and engineering, Xinjiang University, Urumq, Xinjiang, 830000, China

[2] College of Geography and Remote Sensing Sciences, Xinjiang University, Urumqi 830017, China .

[a]liaoyuan@xju.edu.cn, [b]chinayg@xju.edu.cn

Abstract: In the field of weather forecasting, short-term precipitation forecast is still a global problem. The complicated mechanism of convective weather system, the medium and small scale weather system brings great difficulties to the forecast, which is difficult to detect, understand and forecast. The application of short-term precipitation forecast is still in the initial stage. At present, the optical flow method based on radar echo is widely used in short-term and imminent precipitation prediction, but the optical flow method has obvious limitations. In order to improve the forecast precision of short-term and imminent precipitation forecast model, this study introduces the method of depth learning to build the forecast model. The application of depth learning to short-term precipitation forecasting is still in its initial stage. We analyze the existing models may exist in the shortcomings of the proposed in-depth learning model combination of different ideas. By combining different network structures (such as CONVLSTM, CONVGRU, etc.), we can get a better combined forecasting model. It has certain application value.

1. Introduction

Precipitation forecast is an important research topic in meteorology, it is closely related to people's production and life, social economy and so on. The short-term and imminent precipitation forecast is the precipitation weather forecast in 0-6 hours, which is mainly aimed at the medium and small scale weather system and some strong convective weather system. The short-term and imminent precipitation forecast can catch the occurrence of precipitation in advance and accurately predict the precipitation preparation area after a few hours [1].

The temporal resolution and precision of short-term and impending precipitation forecast are higher than those of traditional short-term or medium-term forecast, which is quite challenging, therefore, it has become one of the research hotspots in the field of meteorology.

The traditional methods of short-term and imminent precipitation forecast mainly include statistical forecast method, numerical model forecast method and radar extrapolation forecast method. At present, the short-term and imminent prediction is mostly based on the optical flow method of radar echo, but the optical flow method has some limitations, because the optical flow estimation step and the radar echo extrapolation step are separated, the parameter determination becomes more difficult.

In view of the limitations of the above-mentioned traditional methods for short-term precipitation prediction, in recent years, the theory of depth learning has been rapidly applied to the study of short-term precipitation prediction, it makes the short-term and imminent precipitation forecast based on depth learning become a new research field.

2. Analysis of research status and development trends at home and abroad

2.1 Precipitation and meteorological short-term and impending prediction method based on deep learning

The short-term and impending precipitation forecast based on deep learning is a new field. However, domestic and foreign experts began to explore the research of deep learning and neural network in precipitation prediction very early. Abroad: Ramirez, Velhob and Ferreira analyzed two statistical models of rainfall forecast in Sao Paulo, Brazil. The conclusion is that, compared with multiple linear regression (MLR) model, ANN model has better performance, especially for nonlinear phenomena such as rainfall prediction. B. Root et al. Established an MLP network for precipitation prediction through radar and observation data [2].

The research on precipitation forecast based on deep learning is late in China. In 2015, the public meteorological service center of the China Meteorological Administration and the Institute of automation of the Chinese Academy of Sciences jointly built the "Joint Laboratory of meteorological big data and machine learning"; In 2018, four scientific research institutions including the school of Mathematical Sciences of Peking University jointly established the meteorological big data laboratory. The main research directions of the two joint laboratories include the application of machine learning and deep learning technology to improve the accuracy of precipitation prediction.

In the research of meteorological short-term and imminent forecast based on deep learning, Chen Cheng and Cheng Xingguo proposed a conv Gru model combined with convolutional neural network (CNN) Gru. Since the structure of Gru is simpler than that of LSTM, but the effect is similar, this model has faster training speed and smaller memory requirements than conv LSTM structure. In addition, the convolution layer is improved based on VGGNet. Multiple small convolution kernels are superimposed instead of large convolution kernels, which reduces the number of parameters and improves the feature extraction ability of the network [3]. Hantingting and Zhang Jun proposed to use convolutional neural network (CNN) to build fog short-term and imminent prediction model, and use long-term and short-term memory network (LSTM) to build fog short-term and imminent prediction model. The experimental results proved the effectiveness of the proposed LSTM prediction model [4]. Based on Doppler radar data, Guan Pengzhou et al. Improved the traditional machine learning model and the depth learning algorithm emerging in recent years, put forward a variety of statistical improved models for short-term rainfall prediction, and achieved good prediction results [5].

2.2 Application status of short-term and imminent precipitation forecast

With the gradual deepening of the theory of short-term and imminent precipitation prediction, the research in this field is also being applied to various related industries. At present, more applications are focused on the development of the system and platform. Among the relevant studies abroad, the earliest practical application of artificial neural network to precipitation forecast is neural ware, which developed and designed the artificial neural network weather forecast system in 1987. The prediction system adopts a three-layer neural network model. The prediction system inputs the data of ground pressure field and wind field. The practical application results show that the accuracy of the daily sunny and rainy forecast of the prediction system in June is slightly higher than the operational forecast level of local forecasters [6-8].

At home, Zheng Weicai and others designed short-term and imminent precipitation prediction algorithm and developed short-term and imminent precipitation prediction service phrase application system based on radar 0 - 1 hour short-term and imminent precipitation prediction. Xie Xiaohan and Wang Yong introduced the short-term and imminent integrated analysis and forecast system Inca, which was developed by the Austrian National Meteorological Administration. The system uses the classical correlation motion vector extrapolation method, which is an improvement on the pure extrapolation prediction method. At present, the system has been widely used in many countries and regions. The Meteorological Bureau of Baigou new city developed the Dahongmen business meteorological service platform in 2016 and has been put into use. The platform assists logistics enterprises to reasonably dispatch logistics vehicles according to the meteorological conditions of logistics lines and vehicle positioning system, and helps logistics enterprises to fine manage logistics vehicles and avoid meteorological disaster losses. In 2017, the China Meteorological Administration and Tianjin University jointly developed the national severe convection service product processing system. The system uses image recognition, deep learning and other technologies to quickly and intelligently monitor severe convective weather, which indicates that artificial intelligence technology can effectively improve the early warning ability of short-term approaching forecast [2].

At present, the short-term and impending precipitation forecast has received more attention in the fields of meteorology, logistics and so on. Because the theoretical research is still in the exploratory stage, the practical application of short-term and impending precipitation forecast is not many.

3. Key scientific problems to be solved

(1) Selection and dimensionality reduction of prediction factors

Because there are many factors affecting precipitation, and the correlation between the factors is relatively negative and complex, if these factors are directly substituted into the model as input data, it will increase the calculation difficulty of the model, and also affect the accuracy of the model forecast. Therefore, before training the precipitation model, we should first screen the many influencing factors obtained to expect to select the appropriate prediction factors, and use these factors as the input of the model to carry out precipitation prediction modeling. If these primary predictors are directly used to train the neural network model, it will inevitably increase the complexity of the model and degrade the real-time prediction performance of the model. In order to reduce the network training time, the principal component analysis method is used to reduce the dimension of precipitation prediction factors. It is necessary to ensure the accuracy of model prediction while reducing the dimension of precipitation prediction factors. After PCA dimensionality reduction screening, the final candidate precipitation factors are obtained.

(2) Data preprocessing

The radar echo extrapolation data input is preprocessed by the model. Preprocessing includes data normalization, data clipping and data de-noising. It aims to reduce the interference of irrelevant data, improve the data quality and reduce the amount of calculation. Among them, missing values and abnormal values often exist in radar data or rain gauge data. Aiming at the problem that there are a large number of abnormal data and missing data in precipitation historical data, the data quality control process including data anomaly detection and data breakpoint filling is studied. A variety of anomaly detection methods based on machine learning algorithm are compared; The data filling method based on multiview learning is studied and compared with the traditional interpolation method and the data filling method based on integrated moving average autoregressive model (ARIM).

(3) Construction of prediction model

In fact, the application of deep learning in short-term and impending precipitation forecasting is still in its early stage, and it is not clear how to measure the advantages and disadvantages of these models. Although there have been some better prediction methods based on neural network, each method has certain limitations. By analyzing the possible shortcomings of the existing models, this study puts forward the idea of different combinations of deep learning short-term and imminent precipitation prediction models. By trying to combine different network structures (such as convlstm, convgru, et al.) for experiments, the optimal combination is obtained, and finally a better precipitation prediction model is established.

(4) Prediction accuracy evaluation of the model

Four commonly used indicators in hydrometeorological research, namely, TS score, probability of detection (POD), false to alarm (FAr) and critical success index (CSI), are cited to evaluate the short-term and impending precipitation prediction ability of the prediction model.

4. Technical Route

First, precipitation data are collected in Northern Xinjiang. The research data are taken from the national disaster weather short-term approach prediction system (Swan). The combined reflectance data provided by the system are selected as the precipitation data after conversion by the formula provided by Swan. The hourly observation data in Northern Xinjiang in the past ten years from May to September each year are selected as samples for analysis and processing.

The second is to process the data. The input data of the model need to be preprocessed, including data normalization, data clipping and data de-noising, which aims to educe the interference of irrelevant data, improve the data quality, and reduce the amount of calculation.

Third, the neural network models of different combinations are established respectively, and the optimal combination is obtained through experiments, so as to establish a better short-term and imminent precipitation prediction model.

Fourth, the model is used to predict the short-term and imminent precipitation of samples in Northern Xinjiang to verify its effectiveness; And through experiments comparative analysis is carried out with other algorithms.

Fifth, the accuracy of the short-term and imminent precipitation prediction results using the model is evaluated.

5. Experimental means and key technologies

(1) Partition of prediction objects by systematic clustering method

The stations selected for this study are national observation stations. Because the precipitation situation in the northern Xinjiang region is not exactly the same, directly forecasting the precipitation in the region will not only increase the difficulty of forecasting, but also can not grasp the rainfall in the region, which will affect the accuracy of precipitation forecasting. Therefore, the system clustering method is used to partition the prediction objects into classes with common characteristics, so as to reduce the difficulty of prediction.

(2) Definition of evaluation index

To evaluate the prediction effect of precipitation process, it is usually regarded as a binary event. That is, the forecast results and observations are divided into two cases: rain or no rain. The observed data has precipitation, while the forecast data without precipitation is judged as missing. If there is precipitation in the observation data, but there is no precipitation in the forecast data, it is a wrong forecast. If there is no precipitation in the observation data and prediction results, it is determined that there is no precipitation. Determine whether there is precipitation by setting the threshold. If the precipitation intensity is greater than the set threshold, it is determined that there is precipitation. If the intensity is less than the set threshold, there is no precipitation, as shown in Table 1. TS score is usually used to evaluate the accuracy of short-term and imminent precipitation forecast, and the formula is as follows:

Table 1 Evaluation of precipitation forecast

Reality / Prediction	Yes	No
Yes	NA_K	NC_K
No	NB_K	——

TS score: $$TS_k = \frac{NA_K}{NA_K+NB_K+NC_K} \times 100\% \quad (1)$$

Skill score: $$SS_k = TS_k - TS_k' \quad (2)$$

Underreporting rate: $$PO_k = \frac{NC_K}{NA_K+NC_K} \times 100\% \quad (3)$$

Empty report rate: $$FAR_k = \frac{NB_K}{NA_K+NB_K} \times 100\% \quad (4)$$

Where, NA_K is the number of correct stations (Times) predicted, NB_K is the number of empty stations (Times), NC_k is the number of stations (Times) missed, and ts' is the TS score of numerical forecast or superior guidance forecast.

In addition, the probability of detection (POD), false to alarm (FAr) and critical success index (CSI), which are commonly used in hydrometeorological research, are also used to evaluate the short-term and impending prediction ability of the prediction model. The calculation formula is as follows:

$$POD = H / (H + M) \tag{5}$$

$$FAR = F / (H + F) \tag{6}$$

$$CSI = H / (H + F + M) \tag{7}$$

6. Conclusion

Through analyzing the possible shortcomings of the existing models, this study puts forward the idea of combining different models for deep learning. By trying to combine different network structures (such as convlstm, convgru, et al.) for experiments, a better combined prediction model is obtained. The prediction accuracy of the short-term and imminent precipitation prediction model is improved.

Acknowledgments

This work was financially supported by the Natural Science Foundation of Xinjiang Uygur Autonomous Region (2021D01C080).

References

[1] Li Wei. Research on regional short-term and imminent quantitative precipitation forecasting technology.
[2] Wangting, liuyunpeng, Dongchen. (2019) Summary of short-term and impending precipitation forecasting methods and their applications.
[3] Chen Cheng. Research and application of convolution neural network in short-term and impending meteorological forecast.
[4] Hantingting. Research on short-term and imminent forecast of heavy fog based on deep learning [d] two thousand and nineteen.
[5] Guan Pengzhou, Wang Menghao, Li Qian. (2017) Short term rainfall prediction model based on integrated learning and deep learning [C] // selected papers of the fifth National College Students statistical modeling competition two thousand and seventeen.
[6] Golden Dragon Modeling theory, (2005) method and application of neural network weather forecast [m] Beijing: Meteorological publishing house, 95.
[7] AGOSTINO M. (2005) Sounding-derived indices for neural network based short-term thunderstorm and rainfall forecasts [J]. Atmospheric Research, 83(3): 349-365.
[8] AGOSTINO M. (2005) The Use of Sounding-Derived Indices for a Neural Network Short-Term Thunderstorm Forecast [J]. Weather And Forecasting, 20(8): 896-917.

Based on Wavelet Threshold Denoising-LDA and Bilstm Aircraft Engine Life Prediction

Honglin Li*, Jiaxi Wang

Shenyang Aerospace University, School of Aeronautics and Astronautics, 110136, Shenyang, China

*Corresponding author. Email: (lihonglin0217@163.com)

Abstract—Turbine engines operate in extreme environments and require regular maintenance and replacement. It leads to a lot of human and financial investment, and once the aircraft fails during the navigation process, it will cause irreparable losses. Therefore, it is very necessary to propose an engine life prediction model, which can intervene manually before the engine fails, thereby reducing the probability of an air accident. In order to solve the problem, this paper proposes a wavelet threshold denoising-LDA and Bilstm aero-engine life prediction method, and uses the C-MAPSS data set for simulation prediction. Based on this data set, this paper firstly carried out wavelet threshold denoising to the data set to remove noise, and then used LDA to reduce the dimension to screen out the main features. The processed features are input into the bidirectional LSTM network for prediction. By training the above preprocessed data, BIlstm has better adaptability than the convolutional neural network CNN and LSTM, and the prediction error is greatly reduced.

1. INTRODUCTION

The aero engine acts as the heart of the aircraft, the real-time monitoring of its operating conditions and remaining useful life (RUL), as well as accurate prediction and health management (PHM) of aero-engines are of great significance. Compared with the traditional scheduled maintenance, the maintenance under the predictive model can greatly reduce the maintenance cost and the occurrence of catastrophic accidents while ensuring the improvement of operational reliability. Therefore, it has been widely used in various industrial fields in the world, and has become the current research hotspot.

The current RUL prediction methods are mainly divided into three categories: prediction methods based on physical models, prediction methods based on empirical knowledge, and prediction methods based on data-driven. The traditional physical model prediction method establishes a complex data model based on the existing mechanical model and related principles of mechanical design for life prediction. However, due to the harsh and changeable working environment of the aero-engine, the material and manufacturing process of the aero-engine are not ideal and single [1- 3]. This results in difficulties in establishing mathematical models, too many unknown parameters, large prediction errors, and huge costs. The prediction method based on empirical knowledge mainly relies on industry experts to make predictions according to the existing empirical laws, and the results are subjective and prone to misjudgment. At present, more data-driven prediction methods are used, by collecting or simulating the failure data of aero-engines, and using deep learning combined with failure thresholds to predict the remaining life. Mali et al. used a cyclic neural network for the long-term prediction of machine health status based on competitive learning [4]. Guo Liang et al. proposed an RNN-based rolling bearing health index prediction method. However, RNN has gradient disappearance or gradient descent, which eventually causes the network to fail to update[5]. Yuanet al. proposed a method of long short-term memory neural network for lifetime prediction, which effectively solved the problem of gradient disappearance of RNN[6]. Lu Defeng et al. used principal component analysis and one-dimensional convolutional neural network to predict the life of aero-engines[7]. Compared with traditional RNN and ARIMA, the prediction accuracy is significantly improved. The aero-engine prediction under the bidirectional LSTM neural network proposed by Zeng Huijie has a relatively significant improvement compared with the results of Lstm, RNN, and GRU based on the root mean square error evaluation index[2].

However, the noise in the C-MAPSS data will affect the prediction results of the model to a certain extent. And the C-MAPSS dataset has a high dimension, which leads to a long network training time. Therefore, this paper proposes a method based on wavelet threshold denoising-LDA and Bilstm aero-engine life prediction. The wavelet threshold is used to denoise, and after removing the noise of the data, LDA is carried out to reduce the dimensionality to find the feature with the greatest correlation with RUL, to establish the mapping relationship. Based on this idea, this paper establishes a bidirectional LSTM lifetime prediction model based on data preprocessing and compares the model with traditional CNN and LSTM.

2. THEORETICAL INTRODUCTION

2.1. Wavelet threshold denoising

The essence of wavelet threshold denoising is to suppress the useless part of the signal and enhance the useful part. The process is divided into

(1) decomposition: the decomposition process is to select a wavelet signal for n-layer decomposition.

(2) Thresholding process: it obtains wavelet coefficients after thresholding the decomposed layer coefficients.

(3) Reconstruction: Reconstruct the denoised wavelet coefficients to obtain the denoised signal.

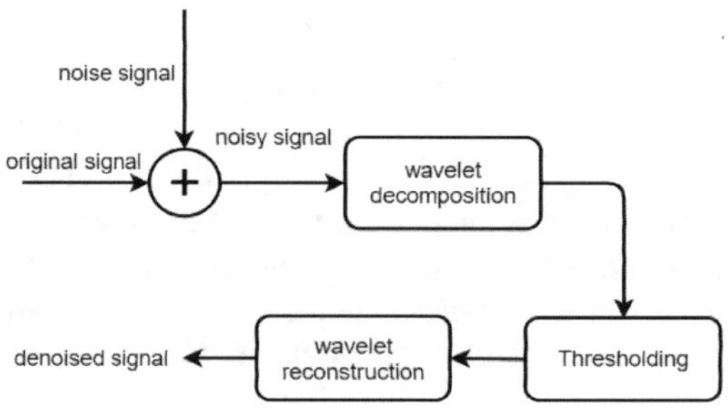

Figure 1 Wavelet threshold denoising process

When processing the signal, it is necessary to select a suitable wavelet base, which should consider the following factors: support length, symmetry, vanishing moment, regularity and similarity. Each feature of the aero-engine data set is the one-dimensional signal, usually, choosing dB wavelet and sym wavelet.

When the signal is decomposed by wavelet, the larger the number of decomposition layers, the more obvious the different characteristics of the noise and the signal, and the better the separation effect. However, if the number of decomposition layers is too large, the reconstructed signal will be more distorted. The relationship between the number of wavelet decomposition layers N and the wavelet frequency range and sampling frequency F_s is as follows:

$$\frac{0.5\times F_s}{2^N} = \frac{f_{max}}{2^N} \tag{1}$$

$$F_s = 2f_{max} \tag{2}$$

f_{max} is the maximum frequency of the signal

In the wavelet domain, the coefficient corresponding to the effective signal is large, while the coefficient corresponding to the noise is small and satisfies the Gaussian distribution. Therefore, the coefficient of the signal in a certain section of the wavelet domain can be set to 0 by setting the threshold value, and the noise can be greatly suppressed. Common threshold selection methods are minimax threshold, fixed threshold, heuristic threshold and unbiased risk estimation threshold. After the threshold method is determined, a threshold function needs to be selected, which includes a soft threshold and a hard threshold.

Hard threshold function: when the absolute value of the wavelet coefficient w is greater than the given threshold thr, the wavelet coefficient remains unchanged; when it is less than the threshold, the wavelet coefficient w_{thr} is set to zero.

$$w_{thr}\begin{cases} w & |w| > thr \\ 0 & |w| < thr \end{cases} \tag{3}$$

Soft threshold function: when the absolute value of the wavelet coefficient is greater than the given threshold, Let the wavelet coefficient subtract the threshold; when it is less than the threshold, the wavelet coefficient is set to 0.

$$w_{thr}\begin{cases} [sgn(w)](|w| - thr) & |w| > thr \\ 0 & |w| < thr \end{cases} \tag{4}$$

2.2. Linear discriminant analysis dimensionality reduction

Linear discriminant analysis (LDA) uses label information reasonably to make the intra-group variance of the same category as small as possible after projection, and the inter-group variance of different categories to be as large as possible, so that the data of different categories can be separated as much as possible.

Given a data set$\{(x_i, y_i)\}_{i=1}^{m}$, the set of examples of the i-th type is X_i, and the mean vector of the examples of the i-th type is μ_i The covariance matrix of the i-th type of examples $\sum_i$, The projection direction is w, the projection of the center of the two types of samples on the straight line: $w^T\mu_0$and $w^T\mu_1$, the covariance of the two types of samples is $w^T \sum_0 w$ and $w^T \sum_1 w$. In order to ensure that the within-group variance of the same category is as small as possible, and the between-group variance of different categories is as large as possible, the following formula needs to be maximized.

$$J = \frac{\|w^T\mu_0 - w^T\mu_1\|_2^2}{w^T \sum_0 w + w^T \sum_1 w} \tag{5}$$

within-class scatter matrix S_w

$$S_w = \textstyle\sum_0 + \sum_1 = \sum_{x\in X_0}(x - \mu_0)(x - \mu_0)^T + \sum_{x\in X_1}(x - \mu_1)(x - \mu_1)^T \tag{6}$$

between-class scatter matrix S_b

$$S_b = (\mu_0 - \mu_1)(\mu_0 - \mu_1)^T \tag{7}$$

Maximize the generalized Rayleigh quotient of the objective function

$$J = \frac{w^T S_b w}{w^T S_w w} \tag{8}$$

Optimize the objective function to find the projection matrix composed of the optimal discriminant vector W_{opt}

$$W_{opt} = argmax\left|\frac{W^T S_b W}{W^T S_w W}\right| = [w_1, w_2, \dots w_n] \tag{9}$$

Using Lagrange multipliers get:

$$S_w^{-1} S_b w = \lambda w \tag{10}$$

Calculate the mean μ_i of each class and the global sample mean μ, and then calculate the intra-class scatter matrix and the inter-class scatter matrix. Perform eigenvalue decomposition on ${S_w}^{-1}S_b$, and take the eigenvector corresponding to the largest eigenvalue. Finally, calculate the projection matrix.

The maximization target corresponds to the maximum eigenvalue of the matrix, and the projection direction is the eigenvector corresponding to the eigenvalue, and then this situation is extended to high dimensions and applied to the model.

2.3. LSTM recurrent neural network

LSTM is an improved recurrent neural network, which can handle time series problems well through the gate structure, while avoiding the problem of gradient disappearance. It is composed of multiple units, and each unit has four components: input, forget gate, output gate and unit state. Its unit composition is shown in Figure 2, C_{t-1} is the state of the previous cell, and C_t is the current updated cell state. h_{t-1} is the output value of the previous state, h_t is the output value of the current cell, and x_t is the input of the current cell. The forgetting gate determines how much information is retained, and its expression is:

$$f_t = \sigma \,(W_i \cdot [h_{t-1}, x_t] + b_f) \tag{11}$$

where f_t is passed to the function by previously hidden information as well as the current input. The input gate determines the state of the cell update and its expression is as follows:

$$i_t = \sigma(W_i \cdot [h_{t-1}, x_t] + b_i) \tag{12}$$

$$\widetilde{C_t} = \tanh\,(W_c \cdot [h_{t-1}, x_t] + b_c) \tag{13}$$

$$C_t = f_t * C_t + i_t * \widetilde{C_t} \tag{14}$$

The output value of the previous state and the input value of the current cell are jointly passed to the sigmoid function, and at the same time, the tanh activation function is passed in to obtain the cell state candidate value $\widetilde{C_t}$ and jointly determine the new state C_t.

The output gate o_t determines the hidden output value of the current state, which determines the information carried by the output gate and the updated cell state. It is expressed by the formula as:

$$o_t = \sigma(W_o \cdot [h_{t-1}, x_t] + b_o) \tag{15}$$

$$h_t = o_t * \tanh\,(C_t) \tag{16}$$

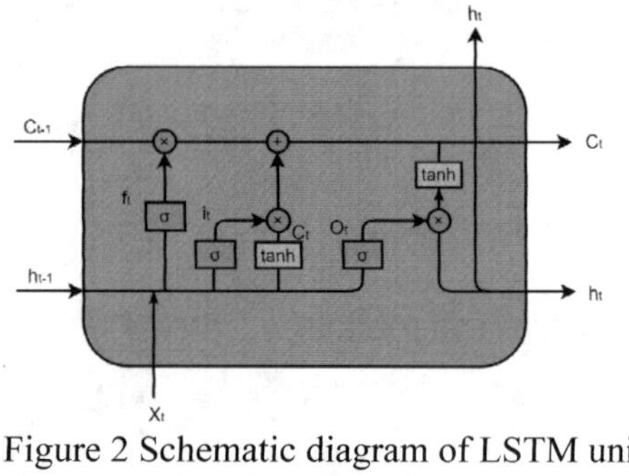

Figure 2 Schematic diagram of LSTM unit

2.4. BILSTM recurrent neural network

Since the LSTM neural network can only make predictions from the information before a certain moment, the bidirectional LSTM neural network is not limited to predicting the result at the moment from the previous information. It can predict the entire input through the time sequence of the input sequence, and divide the hidden layer into two independent hidden layers, positive and negative. The output layer has both past and future features, which increases the accuracy of model prediction. Can get better results than LSTM.

The following diagram is between the bidirectional LSTM structures. The two layers of LSTM read the information of the current time point forward and reverse respectively, and the reversed information will be added with reverse order information. The transmission of hidden information between the LSTM layer and the previous layer not only acts on the adjacent units, but also acts on the input of the next layer. During the training process, since the two state neurons do not interact, it can be expanded into a general feedforward network. Through forward and backward propagation of neurons, the weights of the network can be updated. At present, the LSTM network has greater potential in processing tasks related to timing and logic.

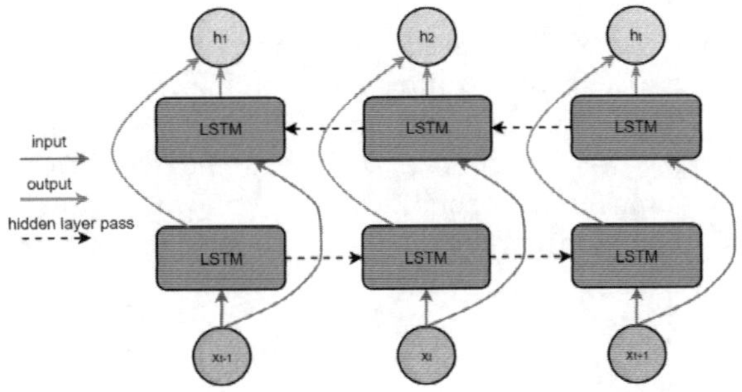

Figure 3 Schematic diagram of Bi-LSTM

3. BASED ON WAVELET THRESHOLD DENOISING-LDA AND BILSTM AIRCRAFT ENGINE LIFE PREDICTION

3.1. Model building process

The overall process of this paper based on wavelet threshold denoising-LDA and Bilstm aero-engine life prediction is shown in the figure. Firstly, the C-MAPSS data set is introduced and then denoised by wavelet threshold to remove part of the noise in the data set. Then, the three operating conditions in the dataset and the 21-dimensional sensor data are reduced in dimension. After experiments, reducing the data to five key features can ensure both the speed of network training and the accuracy of prediction. Then, the denoised and dimensionally reduced data is input into Bilstm for training until the loss function and threshold reach the maximum number of iterations, and then the training is stopped to obtain the model. Finally, input the test set for verification, and use the RMSE, MAPE evaluation index to evaluate the prediction results.

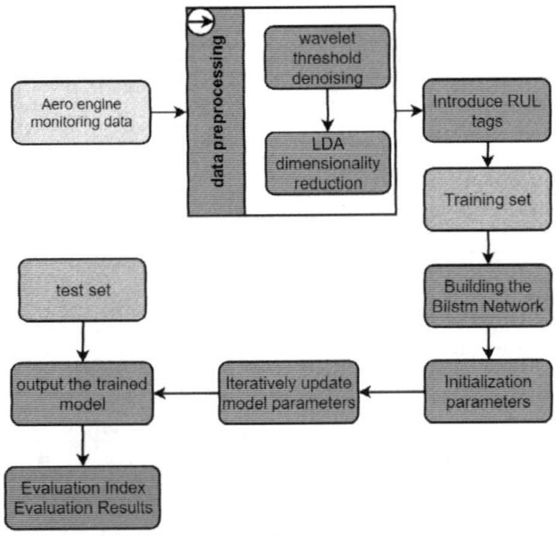

Figure 4 Schematic diagram of the process

3.2. RUL label established

In order to make the model compare the predicted value with the real value during the training process of the aero-engine failure prediction by the bidirectional LSTM neural network of Zeng Huijie[2], it is necessary to add labels. In order to ensure that the system uses the RUL model for prediction after degradation, the piecewise linear degradation model proposed by Ramasso is used for prediction. It is assumed that the lifespan remains constant before a certain stage until the degradation starts when the RUL value is 120, and a relatively good prediction result can be obtained.

4. EMULATION PROOF

4.1. Dataset introduction

This paper quotes NASA's C-MAPSS aero-engine data set for prediction and verification, which is a widely used data set in the field of aero-engine life prediction. It includes four training sets and four data sets, each corresponding to different failure modes and operating conditions. Each data subset contains 26 columns of data, of which the first two columns are the engine numbers and the corresponding time series, then contains three columns of an operational environment, and the last 21 columns are time series data collected by different sensors. The specific description of this dataset can be found in document[8].

4.2. Data preprocessing

Because the data has noise, the widen function in MATLAB is used to denoise the signal before the next processing. This paper uses Stein's unbiased risk estimation principle because of the adaptive threshold. The adjustment process of the threshold is to estimate the noise layer according to the coefficients of the first layer to adjust the threshold. The threshold function in this paper adopts the soft threshold, which performs better when there is more noise than the hard threshold. The basic functions are selected Daubechies wavelets, which are often used to decompose and reconstruct signals and used as filters according to their characteristics. It is abbreviated as dbN, where N is the Daubechies order. After denoising the data, it is found that the denoising effect is the best when the wavelet function order is 4 for three-layer decomposition. The figure below is a comparison image of the No. 11 sensor and No. 15 sensor signal after filtering.

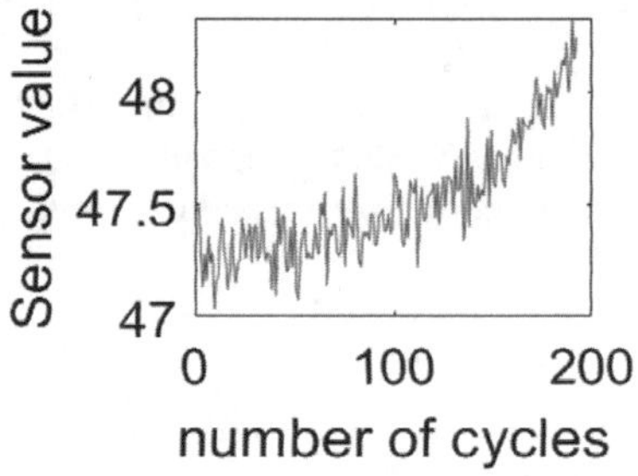

(a) Raw signal of sensor No. 11

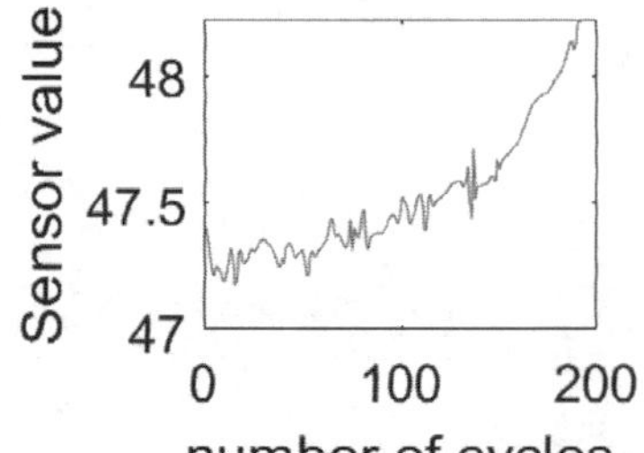

(b) Filter signal of sensor No. 11

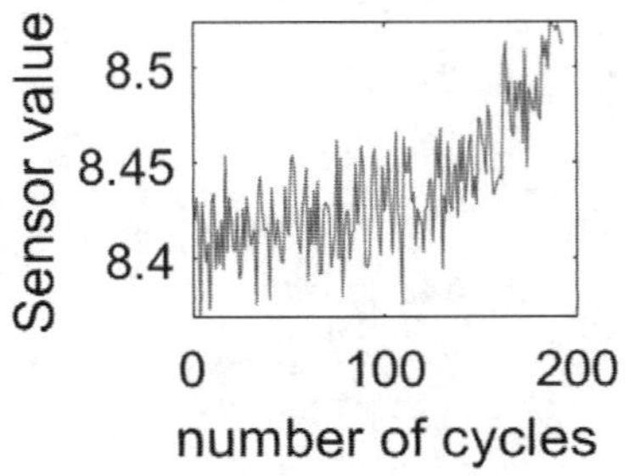

(c) Raw signal of sensor No. 15

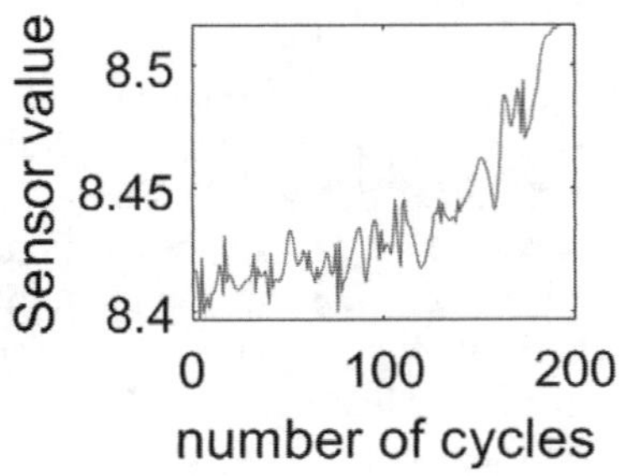

(d) No. 15 sensor filter signal
Figure 5 Signal comparison diagram

Due to the high dimensionality of the denoised data, the network training convergence time will be too long. This paper performs LDA dimensionality reduction processing, and reduces the 24-dimensional features including engine operating conditions to 5-dimensional. Compared with higher or lower dimension selection, the five dimensions not only guarantee no loss of key information, but also increase the training speed. The dimensionality-reduced data can cover most of the data features. Taking the training set FD001 as an example, the features after partial dimensionality reduction are shown in the following table.

TABLE 1 SOME EIGENVALUES

RUL	Fea1	Fea2	Fea3	Fea4	Fea5
97	3.1719e-13	-0.204	-0.0229	-0.4549	-0.1360
77	2.7887e-13	-0.193	0.0872	0.7199	0.1186
32	2.8978e-13	0.7623	0.1260	1.3254	0.4134
16	2.587e-13	1.1385	0.2774	2.1437	0.6735
6	2.0283e-13	1.5228	0.3301	3.5177	0.8603

4.3. Evaluation standard

The evaluation indicators of prediction models mainly include root mean square error (RMSE), mean deviation (MB), normalized deviation (MNB), and mean absolute percentage error (MAPE). In this paper, the root means square error and the average absolute percentage error are used to evaluate the error of the prediction model, and an asymmetric evaluation index is proposed in the paper [8], because the conservative prediction of aero-engine will increase the operating cost but the former is better than the latter when compared with the catastrophic disaster caused by the delayed forecast. So conservative predictions are subject to lower evaluation penalties, and vice versa. Its evaluation function is expressed as:

$$S = \frac{1}{N}\sum_{t=1}^{N} s_t \tag{17}$$

$$s_t = \begin{cases} e^{-\frac{d_t}{13}} - 1, d_t < 0 \\ e^{\frac{d_t}{10}} - 1, d_t \geq 0 \end{cases} \quad (18)$$

In the above formula, N is the total number of samples in the test set; $d_t = p_t - y_t$ is the predicted residual life error; p_t is the predicted value and y_t is the true value.

Percent absolute error expression:

$$MAPE = \frac{1}{n}\sum_{t=1}^{n}\left|\frac{d_t}{y_t}\right| \times 100\% \quad (19)$$

4.4. Emulation proof

This paper first uses the training set to compare the three methods of BiLstm, CNN, and Lstm, and brings the training set data into the trained model to compare the rating indicators of the three methods. In order to improve the accuracy and training speed of the model training, we use data normalization for the features after dimension reduction to ensure that the measurement data can be between [0, 1]. Among them, the number of BiLstm hidden layer nodes is 100, the maximum number of iterations is 300, the learning rate is 0.01, and the gradient is set to 1. The maximum number of iterations of Lstm is 300, the initial learning rate is 0.01, and the miniBatchSize is set to 20. The maximum number of CNN iterations is 20, the gradient threshold is 1, and the initial learning rate is 0.005. For the same dataset, the prediction effects of the three models are as follows.

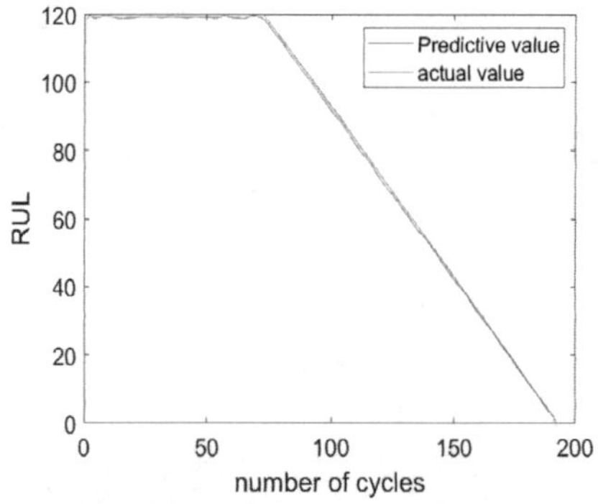

(a) BiLstm lifetime prediction

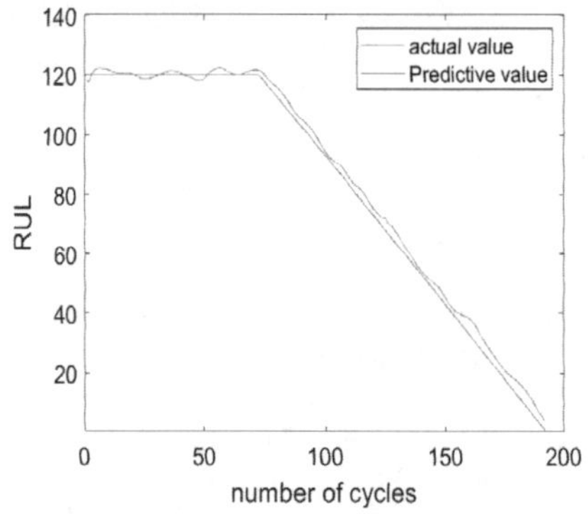

(b) LSTM lifetime prediction

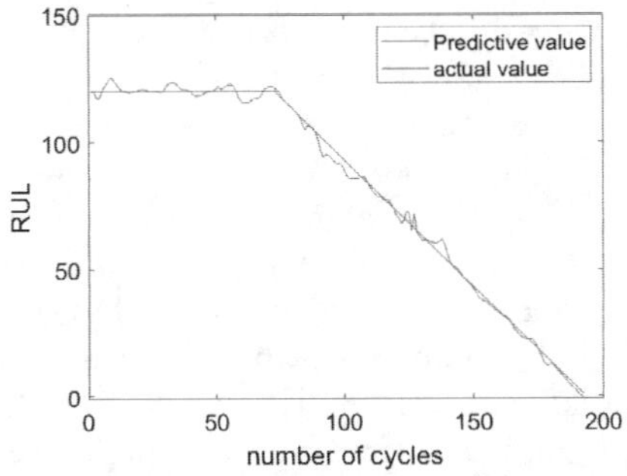

(c) CNN lifetime prediction

Figure 6 Dataset prediction comparison

The above three scores are shown in the table:

Table 2 COMPARISON OF EVALUATION INDICATORS

Model	Evaluation indicators	SCORE
Bi-Lstm	RMSE	0.8259
	MAPE	0.0180
Lstm	RMSE	1.4260
	MAPE	0.0504
CNN	RMSE	2.3657
	MAPE	0.0545

Among them, CNN has the worst effect and is mostly used in the field of image processing. It is not sensitive to one-dimensional data. The effect of LSTM is better than that of a convolutional neural network. It is found that bidirectional Lstm has better adaptability. Therefore, this paper uses 100 samples for bidirectional Lstm. After training with 80% of the data, use 20% of the data for prediction, and the effect is as follows: RMSE: 5.1347, MAPE: 0.2558.

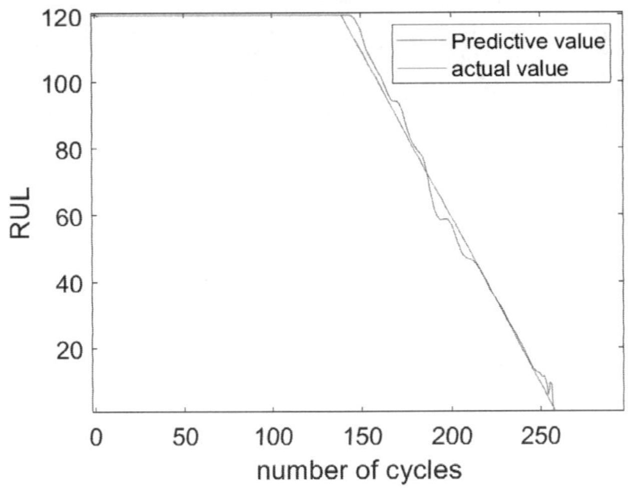

Figure 7 BiLstm prediction

5. CONCLUSION

This paper proposes data prediction based on wavelet threshold denoising and LDA dimensionality reduction processing on the C-MAPSS data set. By comparing the three models of Bi-Lstm, Lstm and CNN, it is found that bidirectional LSTM is more suitable for the problem of aero-engine RUL prediction.

Compared with Lstm and CNN, Bi-Lstm converges faster in the prediction process and achieves better prediction results.

On the other hand, in data preprocessing, the error on the test set is smaller and the network training speed is faster by comparing with the data without wavelet threshold denoising and LDA dimensionality reduction. Based on this idea, it can be known that there is noise in the data set and proper dimensionality reduction can help the algorithm to converge faster and directly affect the operation speed.

In the future, by introducing wavelet threshold denoising, LDA dimensionality reduction processing and more data support and continuously improving the bidirectional LSTM network structure or adjusting training parameters, the prediction of aero-engine RUL can be more accurate, both in civil aviation and in military aviation. The repair and maintenance of turbine engines reduce operating costs and reduces the human, material and financial investment due to real-time monitoring.

REFERENCES

[1] Zhu Yubin; Fang Zhongxiang; Hou Chaohuan; Ma Xiaochuan, Research on aero-engine fault prediction technology based on component parameter estimation, [J], Measurement and Control Technology, 2012, 133-136, 133-136

[2] Zeng Huijie;Guo Jiansheng.Aero-engine fault prediction based on bidirectional LSTM neural network[J].Journal of Air Force Engineering University(Natural Science Edition), 2019, 20(04): 26-32.

[3] Li Jingfeng;Chen Yunxiang;Xiang Huachun;Cai Zhongyi.Remaining life prediction of aero-engine based on LSTM-DBN[J].Systems Engineering and Electronic Technology,2020,42(07):1637-1644.

[4] MALHI A,YAN R, GAORX.Prognosis of Defect Propagation Based on Recurrent Neural Networks[J].IEEE Transactions on Instrumentation & Measurement,2011,60(3):703-711.

[5] GUO L, LI N,JIA F,et al.A Recurrent Neural Network Based Health Indicator for Remaining Useful Life Prediction of Bearings[J]. Neurocomputing,2017,240:98-109.

[6] YUAN M ,WU Y ,LIN L .Fault Diagnosis and Remaining Useful Life Estimation of Aero Engine Using LSTM Neural Network[C]//2016 IEEE International Conference on Aircraft Utility Systems(AUS).Beijing:IEEE,2016:135-140.

[7] Lu Defeng;Hu Yuwen,Remaining life prediction of aero-engine based on principal component analysis and one-dimensional convolutional neural network (English),[J],Transactions of Nanjing University of Aeronautics and Astronautics,2021,159-167,159-167

[8] SAXENA A,GOEBEL K, SIMON D,el al.Damage Propagation Modeling for Aircraft Engine Run-to-failure Simulation[C]//2008 International Conference on Prognostics and Health Management. Denver, CO,USA:IEEE,2008:1-9.

[9] Liu Changyuan; He Xianping; Yu Huiyue, Data-driven residual life prediction of turbine engine, [J], Journal of Electric Machinery and Control, 2021, 72-78, 72-78

[10] Lv Qiuxia;Zhong Xiaoqing;Ren Yasi,Guangzhou-Zhuhai Intercity Short-term Passenger Flow Prediction Method Based on Bi-LSTM Network,[J],Journal of Wuyi University(Natural Science Edition),2022,54-60,54-60

[11] Wang Zhuo;Zhu Ningning;Zheng Xiang,Research on partial discharge pattern recognition method of switchgear based on LDA and RBF neural network,[J],Electronic Measurement Technology,2021,154-158,154-158

[12] Ma Jiancang;Liu Xiaolong;Chen Jing,The combination of second-generation wavelet noise reduction and blind separation applied to aero-engine vibration signal analysis,[J],Mechanical Science and Technology,2010,13-17,13-17

Design and Analysis of Upper Limb Passive Exoskeleton Robot for Lift Assist

Yeyan Ning[1*], Yunji Li[1], Zhenyu Zhang[1], Jun Cai[1], Chunlei Wang[1]

[1] Robot Center, The 21st Research Institute of China Electronics Technology Group Corporation, Shanghai, Shanghai, 200233, China

[*]Corresponding author's e-mail: ningyeyan@126.com

Abstract. Exoskeleton robots are developing rapidly in the field of industrial manufacturing and medical rehabilitation as an intelligent wearable power assist device, which can reduce the lifting burden and muscle fatigue. Existing upper limb passive exoskeleton robots still need to be improved for human-machine integration. In this paper, by establishing a human kinematic model and a lumbar spine motion model, the lifting assist is analyzed, and the design principle of human-machine integration is used to complete the passive upper limb. In structural design, the exoskeleton robot is divided into 4 structural modules, combined with the specific motion form of the spring energy storage module to complete the cam curve design, and the EMG test system is used to collect the surface EMG of the user without wearing and wearing the exoskeleton robot. According to the signal EMG data, the experiment shows that the amplitude of the muscle EMG signal in the wearing state is smaller than that of the non-wearing muscle EMG signal, and the amplitude of the muscle activation amplitude varies from 0 to 10%, which verifies the effectiveness of the upper limb passive exoskeleton.

1. Introduction

In the field of robotics, the market demand has been strong in recent years, and the development prospects are broad. Exoskeleton robots are developing rapidly in the field of industrial manufacturing and medical rehabilitation as an intelligent wearable power assist device [1-3]. In the assembly, welding, painting, and logistics workshops of automobile production lines, high-intensity and repetitive arm lifting actions are very easy to cause injury to workers' arm muscles and reduce production efficiency. The wearable upper limb exoskeleton robot provides a reliable solution for lifting assist. It realizes the follow-up assistance to the human arm by means of spring energy storage, and balances the weight of the user's arm without external energy support, only relying on mechanical devices. The flexibility of the structure itself realizes the lift-assist action, which has good controllability and flexibility, reduces muscle strain and shoulder fatigue, reduces the burden on the body, and provides lift-assist, exercise assistance, body protection and other functions [4-6].

Exoskeleton robots can be divided into two types: active assistance and passive assistance. Active power assistance means using external energy as a power source, including electricity, pneumatic, hydraulic, etc. [7-9]; passive power assistance means that it does not rely on unnatural external forces for assistance, and mainly relies on energy storage structures such as springs and torsion springs to achieve power assist [10-12]. Passive exoskeleton robots get rid of the dependence on the energy system and only rely on the design reliability of the mechanical structure to achieve energy storage and release. It is also more guaranteed, and at the same time, it can provide personalized design and customization according to the specific assistance needs of users, to realize the needs for personalized, comfortable and flexible assistance.

Wang[13] proposed a new wearable structure that optimized the shoulder strap connection configuration and suit layout to improve the excessive shear force of the traditional wearable structure design; Yin [14] proposed the PULE upper limb exoskeleton using a passive suspension system, Tension spring-driven actuators can generate different support forces at different heights to support the wearer's arm; Stienen[15] proposed that the Liimpact exoskeleton robot used a passive damping design, and proposed a rotational design of hydro-elastic actuators (rHEAs), which customized the fit Rotating hydraulic actuators and symmetrical torsion springs as well as enhanced the load capacity of the upper limb passive exoskeleton robot.

The research and development of upper limb passive exoskeleton robots involve the integration of multiple disciplines such as mechanical structure, binding system, human kinematics, human biomechanics and materials science. Starting from the human kinematics model, this paper analyzes the lift assist process, fully considers the factors of human-machine compatibility and completes the design of the upper limb passive exoskeleton robot for lift assist. It analyzes the design principles and design ideas of each module, and renders the upper limb passive exoskeleton robot with a lightweight, flexible and convenient design.

2. Human kinematics model analysis

2.1. Human body model

The source of various movements of the human body is the skeletal muscle system, and the changes in the skeletal muscle system can be reflected by the kinematic model. The design of the upper limb passive exoskeleton robot needs to analyze the change of the motion model of the human body during the lifting process as the design basis. Since the human body is closely related to the spatial position and angle changes of the trunk and limbs when performing the lifting state, it is necessary to analyze the basic characteristics of human biomechanics and anatomy, to reproduce the movement mechanism of human skeletal muscles, and to assist in the establishment of the lifting dynamic model.

As shown in Figure 1, the user's body can be divided into three mutually perpendicular cross-sections: 1-sagittal plane, 2-coronal plane, and 3-horizontal plane according to the axis, and the user performs the lifting action in the sagittal plane.

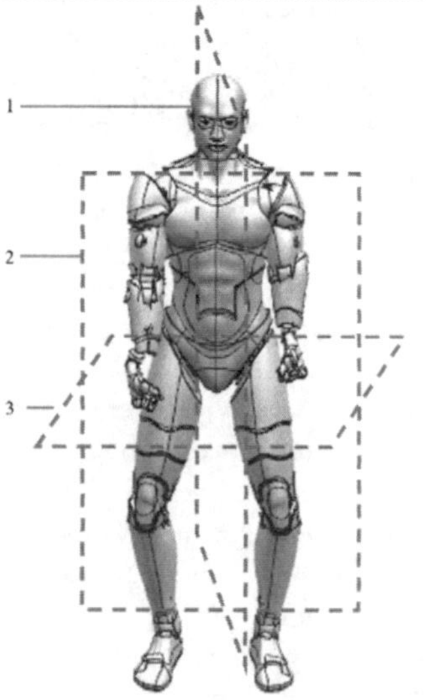

Figure 1. Schematic diagram of the human body.

When the lift assist is realized, the human body is mainly borne by the arms and the lumbar spine.

2.2. *Lumbar spine exercise model*

When the human body performs the lifting action, the spine moves within a certain range with the bending or straightening action. The range of motion of the spine is determined by five parts: coccyx, iliac, thoracic, cervical and lumbar spine. The larger range of motion is the Cervical and lumbar spine.

The spine is under pressure from several aspects: (1) The upper part of the operator's own gravity will produce a shear force on the spine and increase the load on the spine; (2) When lifting heavier objects, the load on the spine of the carrier will be It is related to the distance from the object to the body. The distance between the center of gravity of the object and the spine affects the length of the moment arm, which in turn affects the load on the spine.

The load analysis of the spine is the dynamic force analysis of the body under lifting.

The lumbar vertebra is the main function of the operator's entire bending movement. Since the lumbar vertebra is at the lowest position of the spine and is at the connection between the fixed segment of the spine and the movable segment of the spine, the main force is on the lumbar vertebrae joints.

The lifting state is that the upper body is tilted straight and the legs are transported, and the lifting method is the back lifting type. The entire lifting action of the operator is mainly played by the arms, and the lumbar spine plays an auxiliary role. The lifting action is mainly composed of arm lifting and When the lowering is completed, the lumbar vertebra is flexed forward during the lifting. The bending, stretching and lateral flexing of the lumbar vertebra can produce tensile and compressive stress on the intervertebral disc. 45°.

As shown in Figure 2, the lumbar spine is divided into five parts. The range of flexion and extension from top to bottom shows an increasing trend, which is named L1, L2, L3, L4, and L5. Among them, the L5 part has the largest range of flexion and extension and bears the load. The largest, the lumbar spine of the L5 segment is used as a dynamic model for analysis, which is more representative.

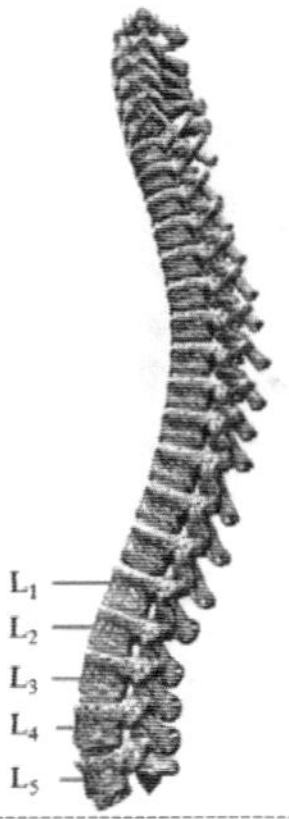

Figure 2. Schematic diagram of the human body.

The forces acting on the lumbar spine are divided into torsion, shear, bending and compression. Among them, the torsional force and the bending force are caused by the bending or twisting of the spine around the center of rotation, the shear force is the component force parallel to the intervertebral disc, and the pressure is the component force acting vertically on the horizontal section of the spine.

3. Upper limb passive structure design and analysis

3.1. *Upper limb passive structure design*

The upper limb passive exoskeleton robot includes an energy storage module, a link transmission module, a back support module, a toothed spine module, and a lumbar support module, as shown in Figure 3.

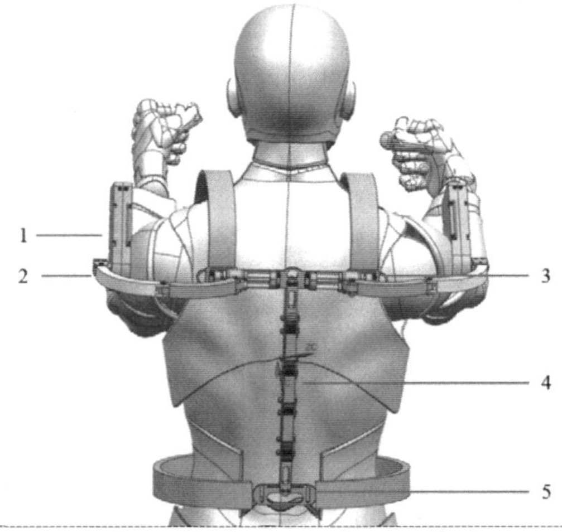

Figure 3. Schematic diagram of upper limb structure.

The energy storage module includes cams, springs, guide shafts and other matching components. The energy storage and energy release process is as follows: the shell is connected to the fixed block above the spring. The initial state is that the shell and the horizontal plane are vertical, and the energy storage module is bound. With the outside of the human deltoid, the arm rotates in the sagittal plane to drive the energy storage module to rotate. During the rotation of the energy storage module, the rotation of the bearing around the cam is converted into a linear motion through the cam roller bearing, and the spring compression energy storage and energy release are completed. , when the arm goes up, the bearing moves reversely from the far rest to the near rest, and the spring releases energy at this time; when the arm goes down, the bearing moves from the near rest to the far rest, and the spring is compressed at this time, and the kinetic energy generated by the arm swinging downward is stored.

The spring energy storage module is placed on the shoulder, which completes the spring compression and extension process through the change of the cam lift. The spring is compressed during the arm swing, and the spring is released during the arm lift process, simulating the lifting and releasing process of human muscles. The force changes in the middle, so that the energy when the human arm is released downward is stored and utilized, and the energy is released when the arm is lifted upwards to assist in the lifting action, reduce muscle fatigue, and achieve the purpose of reasonable assistance. The arm binding strap installs the spring energy storage module on the outside of the deltoid muscle of the arm through the strap mounting piece, which effectively prevents the spring energy storage module from shifting downward due to inertia.

The connecting rod transmission module includes a connecting rod and a rotationally connected bearing assembly. The connecting rod transmission module has two rotational degrees of freedom, which ensures that the upper limb passive exoskeleton robot can achieve a free rotation of nearly 150° in the horizontal plane. The connecting rod transmission module can meet the needs of users of different heights and weights and can adjust the size of the wearable.

The back support module includes the back support plate, the slider rail structure, the toothed spine structure and the lumbar support structure, the back support plate

It is fitted with the back and the weight of the overall upper limb passive exoskeleton robot is transferred to the waist. The slider-rail structure provides the freedom of horizontal movement on the horizontal plane, and the slider moves on the linear guide rail, which can provide a moving distance of nearly 80mm.

The toothed spine module has 4 rotational degrees of freedom and is fixedly connected with the back support plate and the lumbar support structure. While transmitting the force, the spine structure can bend with the bending of the human body. The toothed spine structure is divided into five segments, which are distributed with the human spine. The same is more suitable for human needs.

The waist support module is bound to the waist of the human body through straps, and the overall structural weight is shared with the waist through the waist support module, reducing the load-bearing burden of the upper limb exoskeleton on the human body. The back support structure includes slider guide rails and support rods, which effectively disperse the weight of the overall upper limb passive exoskeleton robot to the waist, reducing the load on the arms and shoulders. The shoulder support module and the waist support module fit the shoulders and waist of the human body, effectively improving the stability of the overall passive exoskeleton robot structure.

The upper limb passive exoskeleton robot has a total of seven rotational degrees of freedom, which can realize the free movement of the human arm in the sagittal and horizontal plane, as well as bending rotation in the coronal plane. The degree of freedom in the sagittal plane is provided by the energy storage module, which has one rotational degree of freedom. The arm can drive the energy storage module bound to the outside of the arm to achieve 360-degree rotation in the sagittal plane, and the degree of freedom in the horizontal plane is provided by the link transmission module, with two degrees of freedom of rotation, when the arm is perpendicular to the body, it can be freely stretched inward and outward.

The overall structure of the passive exoskeleton robot is similar to that of the joints of the human body. It will not collide and interfere with the human body during the movement, nor will it hinder the movement of the user. The passive power assist method adopted reduces the structure in the design concept. The complexity of the robot reduces the overall weight of the robot and increases the wearing comfort. The exoskeleton structure balances the requirements of lightweight and compliant transmission. The compliant structure design can reasonably store and use the human body's own motion consumption, and reasonably transmit it to the required assist position through the mechanical structure to realize the coordination between the user and the passive exoskeleton. The lightweight design ensures freedom of the movement and the stability of power output.

3.2. Cam Curve Design

As the working principle of the spring energy storage module is to convert the rotation around the tangential direction of the cam into linear motion to realize the compression and release of the spring. To increase the distance of the roller bearing when it rotates around the tangent line of the cam profile, the cam is designed as an offset roller push rod linear cam, and the motion law of the driven cam roller in the thrust stroke is sinusoidal acceleration. The shock suffered is a flexible shock, and the distance of the cam lift range is within 70% of the limit compression distance of the selected spring, ensuring that the spring will not fail due to compression reaching the limit range. The cam curve is drawn in matlab, and its curve is shown in Figure 4.

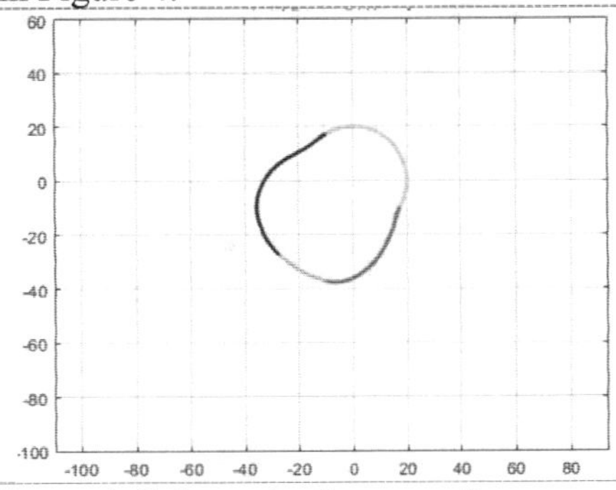

Figure 4. Schematic diagram of cam curve.

The four different colors in the figure represent the four movement stages of the cam, cyan is the near rest stage, blue is the push stage, green is the far rest stage, and red is the return stage. The spring compresses the same distance. In the initial state, the spring energy storage module is vertical to the horizontal plane. At this time, the roller bearing is located near the rest of the cam. At this time, the spring energy storage module is bound to the arm through the strap, and the supply provided by the arm itself is provided when it rotates downward. The energy moves the bearing from the near rest from the lift end to the far rest, completing the spring energy storage, and when the arm is lifted up, the roller bearing moves from the far rest to the near rest, releasing the spring energy.

4. Upper limb passive experimental test

By using the user wearing the upper limb passive exoskeleton robot under the experimental conditions, collecting the surface electromyographic signal EMG data of the experimenter in the state without wearing the exoskeleton robot and wearing the exoskeleton robot, it is verified that the upper limb passive exoskeleton robot can help lift and reduce the lifting force. Effectiveness of fatigue. The instrument in this experiment is the Delsys Lab all-wireless surface electromyography test system. The sensor of the device directly transmits data wirelessly, and the tester does not need to carry a receiver other than the sensor; the EMG sensor and the three-axis acceleration sensor are fused on the same sensor, which can provide a reliable data combination. The experimental system is shown in Figure 5.

Figure 5. EMG experimental system.

In this experiment, the EMG signal on the surface of the human body reflects the strength of each muscle group of the upper limb when the human body is lifting an object. Therefore, it is necessary to determine the position of the muscle group that exerts a force when the human body is lifting an object, and paste it on the position. On the wireless EMG signal transceiver. The sticking position of the upper limb EMG sensor is shown in Figure 6.

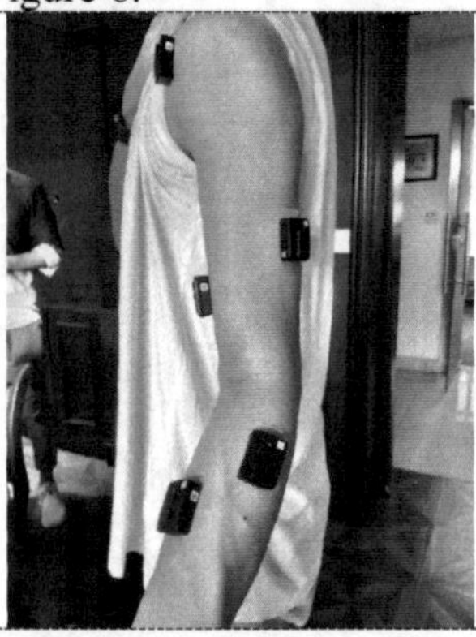

Figure 6. Position of sticking EMG sensor.

The main muscle groups are affected when the human arm is lifted: deltoid, biceps, triceps, flexor carpi ulnaris and brachioradialis. Among them, the deltoid is the most forceful when the arm is lifted, hence it is more representative to analyze the EMG data of deltoid muscle.

The test is carried out in the state of no wearing and wearing the passive exoskeleton robot of the upper limbs, and the test is carried out by lifting a 2KG heavy object, and the EMG sensor data is obtained to estimate the muscle force and muscle activation of the deltoid muscle during lifting, as shown in Figure 7.

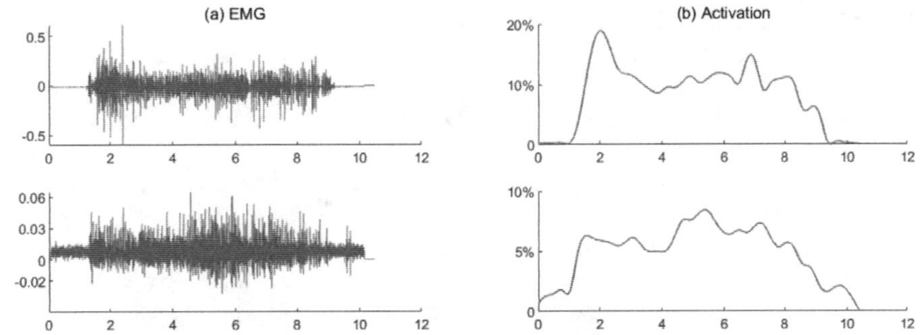

Figure 7. Comparison of EMG Signals and Muscle Activation.

In the non-wearing state, the EMG signal amplitude of the deltoid muscle is ±0.5mv, and the amplitude variation range is 0-20%. When the upper limb passive exoskeleton is worn, the EMG signal amplitude of the deltoid muscle is ±0.1mv. The range of amplitude variation is 0-10%, indicating that the upper limb passive exoskeleton has a certain boosting effect on lifting, which verifies the effectiveness of the upper limb passive exoskeleton structure.

5. Conclusion

In this paper, from the perspective of human-machine compatibility, a passive upper limb exoskeleton robot for lifting assistance is designed, and the cam curve design is completed in combination with the specific form of lifting assistance. The EMG data of the surface EMG signal in the state, the experiment shows that the amplitude of the muscle EMG signal in the wearing state is smaller than that of the non-wearing muscle EMG signal, and the amplitude of the muscle activation amplitude changes in the range of 0-10%, which verifies the upper limb passive exoskeleton's effectiveness. However, there is still a certain deviation between the joint axis of the designed upper limb passive exoskeleton robot and the human body, resulting in human-machine obstruction during movement. The realization of human-machine synergy is the next research direction of passive exoskeleton robots.

References

[1] Ren, Y. , et al. "Developing a Multi-Joint Upper Limb Exoskeleton Robot for Diagnosis, Therapy, and Outcome Evaluation in Neurorehabilitation." IEEE Transactions on Neural Systems & Rehabilitation Engineering 21.3(2013):490-499.

[2] Spada, S. , et al. "Investigation into the Applicability of a Passive Upper-limb Exoskeleton in Automotive Industry." Procedia Manufacturing 11(2017):1255-1262.

[3] Bouteraa, Y. , I. B. Abdallah , and A. Elmogy . "Design and control of an exoskeleton robot with EMG-driven electrical stimulation for upper limb rehabilitation." Industrial Robot: the international journal of robotics research and application 47.4(2020):489-501.

[4] Gao, B. , et al. "Real-Time Evaluation of the Signal Processing of sEMG Used in Limb Exoskeleton Rehabilitation System." Applied Bionics and Biomechanics 2018(2018):1-6.

[5] Wang, W. A. , et al. "Design and verification of a human–robot interaction system for upper limb exoskeleton rehabilitation - ScienceDirect." Medical Engineering & Physics 79(2020):19-25.

[6] Jian, Guo , et al. "Design of the Training Management System for the Exoskeleton Upper Limb Rehabilitation Robot." Life Science Instruments (2017):30-34.

[7] Gopura, Rarc , et al. "Developments in hardware systems of active upper-limb exoskeleton robots: A review." Robotics & Autonomous Systems (2016):203-220.

[8] Wu, Q. , et al. "Development of an RBFN-based neural-fuzzy adaptive control strategy for an upper limb rehabilitation exoskeleton." Mechatronics 53(2018):85-94.

[9] Ouyang, X. , et al. "Development of a novel compact hydraulic power unit for the exoskeleton robot." Mechatronics 38(2016):68-75.

[10] Hu, B. , et al. "Design of Mechanism and Control System for a Lightweight Lower Limb Exoskeleton." International Conference on Control, Robotics and Cybernetics University of Shanghai for Science and Technology; University of Shanghai for Science and Technology; University of Shanghai for Science and Technology; University of Shanghai for Science and Technology; 2018.

[11] Liu, X. , et al. "Adaptive Backstepping Sliding Mode Control for a Hydraulic Knee Exoskeleton Robot." (2020).

[12] Yu, S. , et al. "Design and Control of a High-Torque and Highly Backdrivable Hybrid Soft Exoskeleton for Knee Injury Prevention During Squatting." IEEE IEEE, 2019:1-1.

[13] Wang, J. , et al. "Comfort-Centered Design of a Lightweight and Backdrivable Knee Exoskeleton." (2019).

[14] Yin, P. , et al. "Effects of a passive upper extremity exoskeleton for overhead tasks." Journal of Electromyography and Kinesiology 55.7(2020):102478.

[15] Stienen, Aha , et al. "Design of a rotational hydro-elastic actuator for an active upper-extremity rehabilitation exoskeleton." IEEE Ras & Embs International Conference on Biomedical Robotics & Biomechatronics IEEE, 2008.

Journal of Physics: Conference Series **2410** (2022) 012021 doi:10.1088/1742-6596/2410/1/012021

Research on the Construction Method of Remote Sensing Sensing Information Platform Based on Internet of Things

Hong Zhou*

China National Aero-Technology Import & Export Corporation of Ecommerce, Beijing, China,102600

*zhouhong8878@163.com

Abstract: With the gradual rise and development of Internet of Things technology, the development of related industries has shown new vigor and vitality. From the definition and application of the Internet of Things and remote sensing monitoring, the application fields of the combination of the Internet of Things and remote sensing monitoring are given a specific description. On the basis of analyzing the research status and technology at home and abroad, using VS. NET method to build a three-layer network platform to realize the construction of the information platform, and verify the application of the information platform is better.

1. Introduction

The Internet of Things refers to the Internet where things are connected. It connects any item with the Internet through various information sensing devices to communicate information to achieve intelligent network management. Once the concept of Internet of Things was put forward, it immediately got the attention of academia and industry at home and abroad, and quickly became a very popular research field. In recent years, experts and scholars at home and abroad have carried out in-depth research on the relevant theories of the Internet of Things, and carried out a series of explorations at the practical application level, and achieved a series of results. Remote sensing is a non-contact long-distance detection technology, which monitors and manages the earth's surface by installing remote sensing monitoring instruments on artificial earth satellites. Remote sensing monitoring refers to the use of remote sensing related technologies to monitor various aspects of the ground to provide data and information for people's research in various aspects. Remote sensing monitoring is a large-area monitoring method, which can obtain data information on a large area of the surface at the same time, and is a surface-based monitoring[1-2]. The application fields of remote sensing monitoring are very wide, and now a multi-level, multi-perspective and multi-field observation of detection and monitoring on a global scale has been formed from the ground to the air and even space, from the collection, processing, analysis and application of information data. The system has now become an important way to obtain information on the earth's resources and environment. Although this monitoring method is widely used, this surface-based monitoring method also has its shortcomings, such as weak real-time information feedback. As the main body of the next-generation information network, the Internet of Things will be a complex of many technologies. The key technologies of the Internet of Things include Internet technology, EPC/RFID technology, entity markup language development technology, sensor network technology, wireless communication technology and embedded technology[3-4]. Since remote sensing has the characteristics of large-area synchronous observation, if it is combined with related technologies of the Internet of Things, it will bring great convenience to the development of remote sensing monitoring, and at the same time, it will provide and ensure the timeliness and accuracy of data analysis and processing. The combined application of remote sensing monitoring and Internet of Things technology will make remote sensing monitoring enter a period of rapid development, which can provide a variety of earth observation data information in a timely manner[5-6]. At present, although the technology of combining the Internet of Things and remote sensing monitoring has been applied in many fields and achieved certain results, there are still some problems that have not been solved. Starting from the analysis of the application field of the combination of the Internet of Things and remote sensing monitoring, this paper summarizes the research results of domestic and foreign theory and practice, and lays a good foundation for future in-depth research[7-8].

2. Research status

The technology combining the Internet of Things and remote sensing monitoring is also widely used in people's production and life. In the aspect of logistics, a remote sensing control logistics information system is proposed, which uses intelligent information processing and interactive control methods to provide the next guidance for the real-time update of logistics information in the future. In terms of agricultural production, the technology has been used to monitor the production status and output of crops, indicating that it can monitor the growth status of crops in real time, such as estimating the yield of grain crops. It is based on the crop climate yield prediction model, genetic algorithm combined with neural network model, inversion model based on human-computer interaction, county yield estimation model based on decision tree classification, unit yield estimation model, and CERES based on SCEUA algorithm. The wheat model, radar remote sensing production estimation model and other related model algorithms can accurately estimate production, indicating that is available. The combination of remote sensing monitoring and Internet of Things technology can provide important decisions for improving the accuracy of crop yield estimation; in addition, the prevention and control of pests in the process of agricultural planting also improves the efficiency.

At present, the development of the Internet of Things mainly has the following two problems: the problem of the uniformity of management and technical standards and the problem of technical bottlenecks. At present, the global Internet of Things still lacks a reasonable management organization, and most of the global research on the Internet of Things is still in its infancy. The development of the Internet of Things in China also has these problems: on the one hand, my country lacks the core technology of independent property rights; a series of problems such as few systems; more test systems, less scale and use; the popularization of the public; international cooperation and other issues.

In view of the current problems, the following aspects need to be further improved in the future: Firstly, although the technology of the Internet of Things has been developed from various industries at home and abroad, the most important thing is to establish an Internet of Things management organization. The architecture and usage standards of the Internet of Things are managed uniformly around the world; secondly, in the application of the Internet of Things and remote sensing monitoring, the scope of remote sensing monitoring is also determined by its extensiveness and the degree of fit of the Internet of Things technology, which brings complexity. To solve this problem, the fundamental method is to search and store all kinds of information required for remote sensing monitoring from the bottom layer in detail, so as to facilitate the later call. That is, it is necessary to build from the bottom up, so as to carry out a comprehensive network; thirdly, although the application fields of the combination of the Internet of Things and remote sensing monitoring are very wide, the understanding of the needs of each field needs to be decomposed in detail, and finally established a complete set of facilities.

3. Application of the model in the platform

The design of the system is the foundation of the platform construction. In this study, the B/S structure is used to realize the overall architecture of the system. The specific architecture diagram is shown in Figure 1:

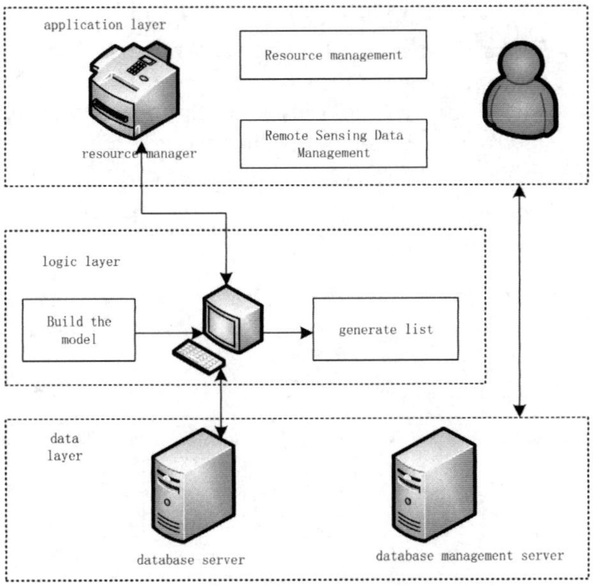

Figure 1 Overall architecture of the system

The design of the information platform should follow the following points:

(1) The platform is easy to operate. Graphical representation is used to display information, which enhances the intuitive sensibility of users logging into the platform for business information, and facilitates the timely processing of business information by staff.

(2) Data structure normalization. The system must meet the data entry format required by business development and on-site practical application, so as to achieve standardized data entry and regular storage.

(3) Intelligent platform analysis. In the research process, the idea of intelligence is adopted, with system intelligence as the main line, and the idea of intelligence is fully reflected in each module of the system.

(4) Platform permission management. Different operators have different rights to use business modules, which prevents the management rights of modules from being crossed during the use of the system, and standardizes the use of module functions.

Development platform/language: VS2010/C#;

Operating system: WINDOWS 7 32-bit;

Database environment: ORACLE 11g;

Description language: UML and XML;

Database environment: ORACLE 11g is used for rule base and system data storage;

Specified protocol: TCP/IP;

System Service Manager: IIS 7.0 and above;

Other applications: Office components;

The basic platform mainly uses the "user-role-service" rights management method to realize the construction of the system basic framework platform. Platform metadata includes basic staff information, authority assignment information, work unit and administrative level and other elements. The platform adopts the accessory customization function to realize the needs of business information customization, and takes the self-adaptive idea as the main idea of the system to flexibly solve the business function problem.

The basic platform collects applications in the business system in the form of services and tools, and allocates modules according to the connection between businesses. Based on the business distribution in actual work, the authority is divided for users. Different users are responsible for different business modules. Only the modules they are responsible for are business management and daily data maintenance. Different permission roles with different work contents are assigned to users, and users with the same permission role have the same module operation and maintenance content.

The resource processor provides the required resources through remote sensing technology to help it quickly obtain the required resource information, which requires a large amount of data as the basis. The interface circuit of the resource processor is shown in Figure 2:

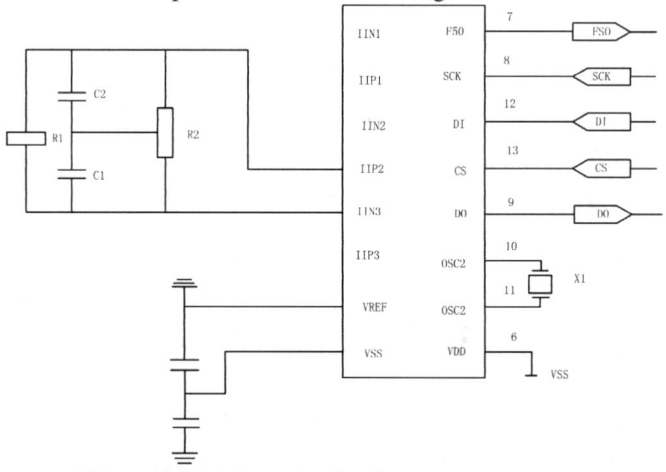

Figure 2 Interface circuit of resource processor

According to Figure 2, in the interface circuit of the resource processor, the nRF905 chip is used to realize the communication function of the resource processor, which not only has super strong transmission capability, but also has high-speed data transmission capability. At the same time, the chip is equipped with 1024 communication channels, which can meet the integration requirements of English teaching video resources. The processing time of its resource processor is ≤6μs, and the working voltage ranges from 1.9V to 3.6V during the working time, and the working current is controlled. Within 12.5μA, it is a low-power working state and can process multiple search requests at the same time.

4. Experimental effect analysis

The data is divided into 10 groups to verify the use effect of the information platform, and to predict whether there is a failure and the use efficiency of the platform, as shown in Table 1.

Table 1 The verification effect of the information platform

Data	Information platform	
	Accuracy (%)	time(ms)
AJ1	83.72	24
AJ2	72.53	29
AJ3	81.16	24
AJ4	77.61	49
AJ5	83.31	15
AJ6	72.91	12
AJ7	79.03	17
AJ8	82.06	23
AJ9	81.12	81
AJ10	79.91	16
AJ11	75.94	15

AJ12	77.23	12
AJ13	81.71	10

Through the verification effect, it can be seen that the information platform proposed in this study can be verified through the overall grouping of the data (the data is randomly divided into 13 groups), which has a good application effect, high platform accuracy and good fault problems, and also has a certain practicability.

5. Conclusion

In recent years, with the importance attached to the Internet of Things technology at home and abroad, after the concept of "smart earth" was put forward, my country has put forward the concept of "smart city", which shows that the development trend of the Internet of Things has increasingly become the focus of national and social construction. Through the analysis of this paper, the related technologies of the Internet of Things can be used to obtain information in real time and conduct on-site verification of remote sensing images. The use of remote sensing monitoring to monitor a large area can make up for the lack of a small monitoring range of the Internet of Things. It can combine points and surfaces to form a state of integration of heaven and earth, and provide real-time and human monitoring work timeliness. At the same time, this study can verify the application effect of the platform through the continuous collection and sorting of data.

References

[1] Zhang Jiaze, Wang Fei, Zhang Shengmao, Zhang Shouyuan. A review of fishery application research based on remote sensing technology [J]. Journal of Zhejiang Ocean University (Natural Science Edition), 2022, 41(01): 81-90.

[2] Yang Ying, Ding Luming, Tang Xia, Zhang Jinlong, Shi Baoguang, Xu You, Li Guozheng. Application of Remote Sensing and Internet of Things Coupling Technology in Dynamic Management of Land Biological Resources [J]. Grass Industry Science, 2021, 38(05): 992-1000.

[3] Gao Demin, Shi Dongxu, Xue Wei, Hu Bin. Research on monitoring technology of agricultural pests and diseases based on Internet of Things and low-altitude remote sensing [J]. Northeast Agricultural Science, 2021, 46(01): 108-113.

[4] Li Kai, Li Feng, Zhang Jianmin, Yang Weiming. Research on the Application and Development Trend of Geospatial Informatics in the Internet of Things Era [J]. Surveying and Mapping Geographic Information, 2020, 45(05): 78-83.

[5] Chen Dongqin, Chang Pengfei, Peng Ahui, Wang Ying. Design of environmental protection supervision system during highway construction based on Internet of Things [J]. Radio Engineering, 2020, 50(04): 280-284.

[6] Zhao Jinling. Analysis of the promotion effect of remote sensing courses on the cultivation of Internet of Things engineering students [J]. Education Modernization, 2019, 6(61): 239-242. DOI:10.16541/j.cnki.2095-8420.2019.61.085 .

[7] Kong Yan, Wang Baoyun. Overview of Remote Sensing Monitoring Based on Internet of Things [J]. Modern Computer (Professional Edition), 2017(19): 38-40+56.

[8] Ren Rundong, Wu Qiulan, Pan Ningning, Liang Yong. Discussion on remote sensing measurement of forest carbon sinks based on Internet of Things [J]. Bulletin of Surveying and Mapping, 2012(S1): 444-446.

Optimal Numbers and Mix of SSA Drones and Radio Repeater Drones based on Ant Colony Algorithm

Shu Yu[1], Haiying Wang[2], Luo Zheng[3,*]

[1] College of Cyber Security, Jinan University, Guangzhou, Guangdong, 510000

[2] Birmingham University joint Institution, Jinan University, Guangzhou, Guangdong, 510000

[3] College of Science and Engineering, Jinan University, Guangzhou, Guangdong, 510000

*Corresponding author. Email: (zhengluo1208@163.com)

Abstract: During the 2019-2020 fire season in Australia, there were destructive wildfires in every state, particularly in Victoria. Therefore, we are required to build a model to deploy drones to surveil the fire. We develop the basic model to find out the optimal numbers and mix of SSA drones and radio repeater drones. First, we use Ant Colony Algorithm to solve the shortest path, which could make the government purchase as few drones as possible under the premise of covering a large area. According to the operating mode of UVA and the theory of UAV formation cooperative control, we find out the information state of the radio repeater and its cruise path length. Besides, we create a fire risk assessment to determine the speed of information updates and adjust the number of drones. Finally, we determine the optimal combination of 1260 SSA drones and 151 repeater drones.

1. Introduction

During the 2019-2020 fire season in Australia, there was a destructive wildfire[1] in every state, particularly in Victoria. Emergency Operations Center (EOC) uses SSA drones and two-way radio communication to monitor site conditions and transmit data and information to surveil the fire and direct the frontline personnel to put out the fire.

However, since their low transmitting power limits the range of handheld radios, EOC hopes to turn to other equipment to extend the range of low-power radios on the front lines. Recently, a company launched a prototype WileE–15.2X hybrid drone. Therefore, EOC could purchase this drone to dramatically extend the range, making information transmission more efficient and faster.

Table 1 WileE–15.2X hybrid drone capabilities

Function	Flight range	Maximum speed	Maximum flight time
Radio repeater, video, telemetry	30 km	20 m/s	2.5 hours

We constructed a model to determine the optimal numbers and mix of SSA drones and Radio Repeater drones to purchase. The model needs to balance capability and safety with economics and consider observational and communications mission needs and topography.

2. Model establishment and solution

2.1. Data processing

We first process the collected satellite remote sensing data and extract the fire data in Victoria. From this, we draw a scatter diagram according to the brightness of the flame, latitude, and longitude. Then, we synthesize it with the topographic map of Victoria to get the Victoria fire distribution map in figure 1.

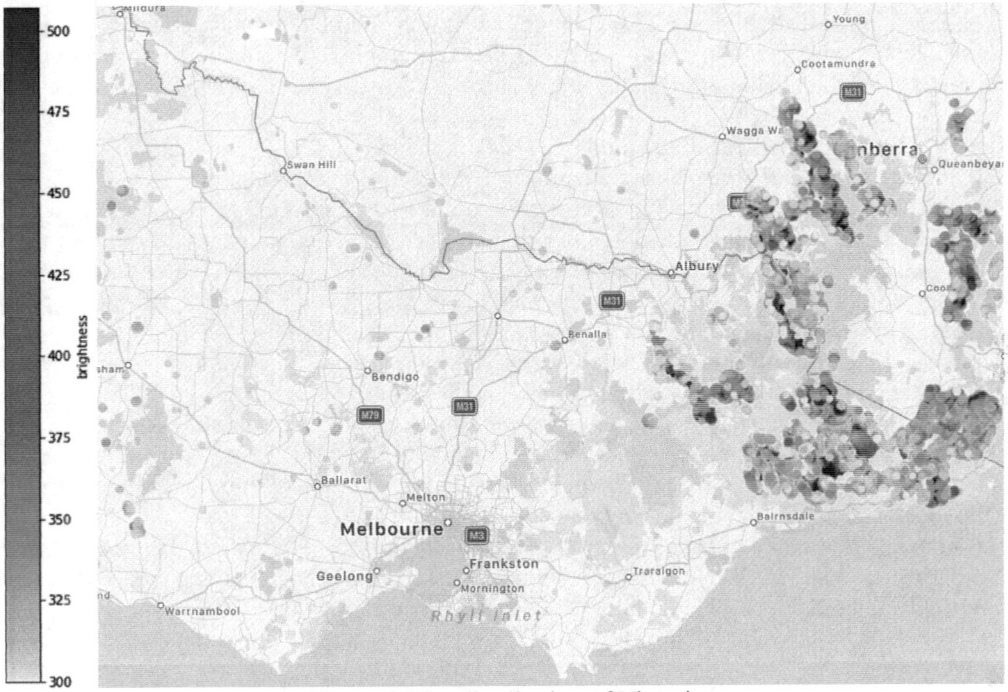

Figure 1 Fire distribution of Victoria

Next, based on terrain and historical fire occurrences, 80 representative latitude and longitude points for three-dimensional mapping are selected to assist in the subsequent calculation of the shortest path and the number of drones. To make the constructed topographic map more in line with the actual situation, we interpolated these 80 data points to construct a topographic map of eastern Victoria in figure 2.

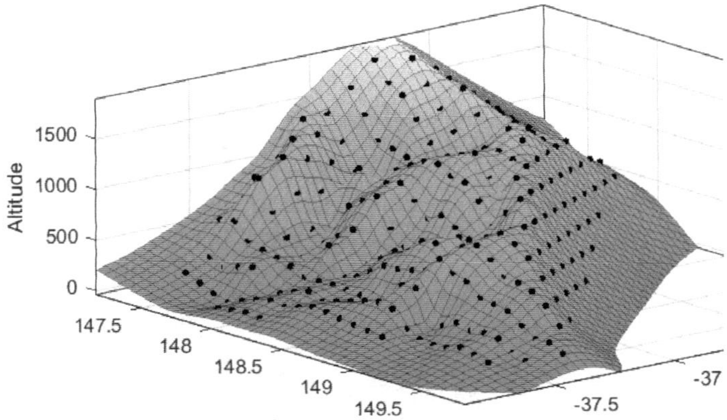

Figure 2 3D Map of Victoria of 80 latitude and longitude points

2.2. Shortest path solving

Based on economic and safety considerations, we need to buy as few drones as possible under the premise of covering a large area. We first apply Ant Colony Optimization (ACO[2]) to calculate the shortest path.

In the initial state, *n* ants are randomly placed in the fire point, the initial setting value of each path is the same, and there are interconnected paths between all fire points. When *n* ants searching among *m* fire points, it is carried out according to the state transition probability, and artificial intelligence ant colony search is more orderly than real-life ant colony search. While constantly choosing the path with high pheromone intensity, *n* ants finally return to the starting point. When the entire road network is searched, the minimum value of all paths will also be recorded simultaneously to optimize the path selection.

Step 1: Initialize the number of iterations. Set N_c=0.

Step 2: Randomly place *n* ants on the location where the fire point will pass.

Step 3: The ants on each node will select the fire point to pass the next time according to a certain probability to complete the traversal of all fire points. Each ant finds its path according to the rules. When the ant *m* is on node *i*, it chooses the next destination node *j* to search according to a certain probability instead of blindly choosing. In this way, ant *m* finds a complete path and finally returns to the starting point to construct a loop.

Step 4: The algorithm statistically compares the paths searched by ant *m* and records the optimal solution generated in the current iteration process.

Step 5: We update the pheromone concentration on each path. After *n* ants' travel, *n* results are produced, and *n* paths are generated. One of these *n* paths is the optimal path. Therefore, the value of each pheromone on the optimal path is updated. The pheromone value of the shortest path of each path is increased, and the algorithm is self-evolved.

Step 6: When $N_c > N_{max}$, the algorithm ends, and the search stops. Otherwise, the loop iteration will continue, that is, $N_c = N_c + 1$. We follow the state transition probability to find the next element. In the process of selecting nodes, the transition probability expression followed is:

$$P_{ij}^k = \begin{cases} \dfrac{[\tau_{ij}(t)]^\alpha [\eta_{ij}(t)]^\beta}{\sum_{k \in allowed_k} [\tau_{ik}(t)]^\alpha [\eta_{ik}(t)]^\beta} & i,j \in allowed_k \\ 0 & \end{cases} \quad (1)$$

Then we compare the shortest distance and average distance of each iteration to illustrate the process of algorithm evolution.

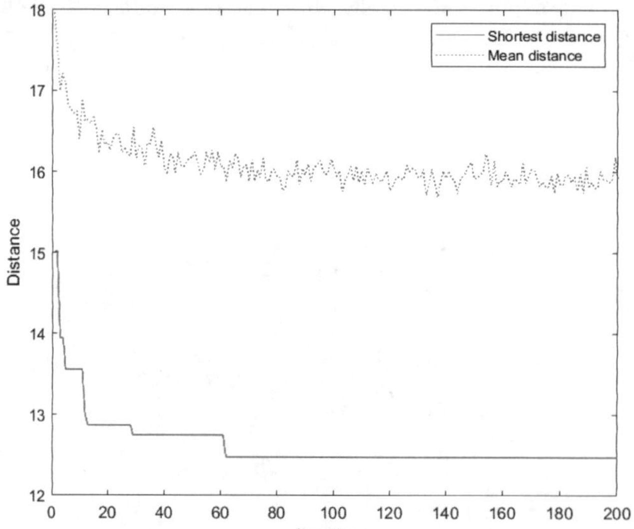

Figure 3 Comparison of the shortest distance and average distance of each generation

In the Ant Colony Optimization, the pheromone must update its state at any time to mark the characteristics of the shortest path. The calculation formula is:

$$\tau_{ij}(t) = (1-\rho)\tau_{ij}(t) + \sum_{k=1}^{n} \tau_{ij}^{k} \quad (2)$$

$$\Delta\tau_{ij}^{k} = \begin{cases} C_k^{-1} & (i,j) \in L^k \\ 0 & others \end{cases} \quad (3)$$

We can obtain the shortest path when the number of iterations N is greater than 62 after a program operation. The shortest path is 12.2882 (latitude and longitude). After transforming, the total is 1363.9902 km.

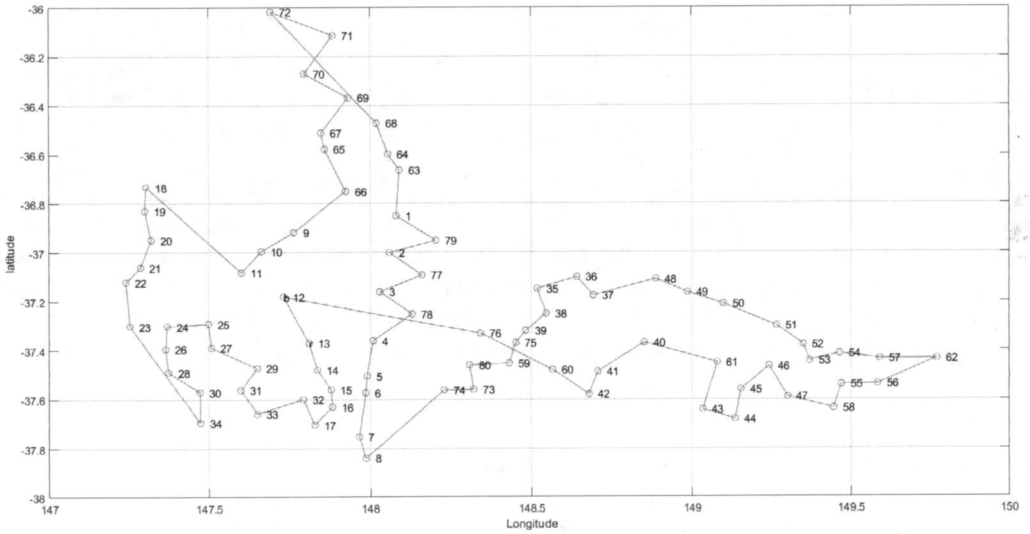

Figure 4 Optimization of path using ACO

2.3. Drone working mode

The rapid wildfire response system consists of the frontline information system, the extended relay system, and the emergency operations centre. Since UAV can be used as a frontline information system and an extended relay system, it mainly affects the first two. Based on the UAV formation cooperative control theory, we should first consider expanding the relay system because this is the bridge connecting the front and rear and the direct controller of the frontline system. It can be regarded as the skeleton of the entire system, and each frontline information device is the node at the end of the skeleton.

Here we have to mention a concept called the information state. Because the operating mode of the repeater is the inspection mode, each area will have an information state, and each time it passes through an area, it will contact the SSA in the area to update[3] the information status in the area, which means the information update is phased. We can intuitively see the change of the regional information status during the flight of the repeater from figure 5.

Since the flight range of each repeater drone is limited, the information area they can cover is also limited. Based on this, we can divide each area according to the shortest path obtained above, and for each area, we can deploy a basic repeater for information inspection and multiple SSA drones for frontline information collection. Limited by the flying distance and range of the UAV, we approximately calculate the maximum coverage area of a single relay UAV as follow: $S=\pi R^2+2RL$.

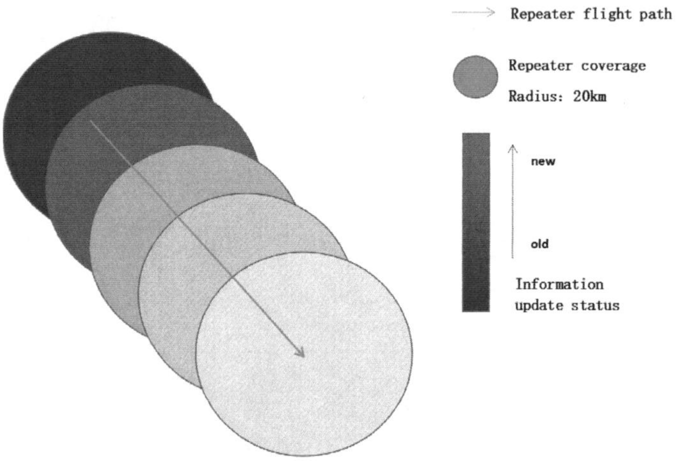

Figure 5 The change of the regional information status during the flight of the repeater

The setting of the cruise path length of the relay UAV is related to the safety and economy of the system, so the calculation of L is calculated by a safety and economic balance coefficient q and the UAV's cruising range l, that is, $L=q\text{-}l$. Besides, the safety economic balance coefficient is a value of [0,1]. The closer it is to one, the more economical it is, but the security risk increases. On the contrary, the closer it is to zero, the safer it is, but the cost and efficiency will increase. After combining relevant papers and referring to Victoria's topography[4], economic conditions, and other factors, we choose q=0.8.

2.4. *Fire risk assessment system*

Since the possibility of fire in different places is not the same, if we choose the same combination strategy for each area, then the area with a lower probability of fire will cause a waste of resources and resources of the region with a high probability of fire may be insufficient. Therefore, we need to establish a fire risk assessment system for risk division, and then we can use it to determine the speed of information update according to the division level to adjust the number of drones.

Here, we mainly consider two aspects: the size of the fire and the frequency of historical fires. The influence of the former can be adjusted immediately, while the latter influences the basic adjustment of the evaluation system. Therefore, we first evaluate these two aspects.

Table 2 Evaluation based on the size of fire

Size of Fire	Score Interval
Particularly serious	$7.5 < P_1 \leq 10$
Serious	$5 < P_1 \leq 7.5$
Large	$2.5 < P_1 \leq 5$
General	$0 \leq P_1 \leq 2.5$
None	0

Table 3 Evaluation based on historical frequency

Historical Frequency	Score Interval
Frequently	$7.5 < P_2 \leq 10$
Usually	$5 < P_2 \leq 7.5$
General	$2.5 < P_2 \leq 5$
Seldom	$0 \leq P_2 \leq 2.5$
Never	0

According to Table 2 and Table 3, we could obtain a final score $P_{sum}=P_1+P_2$. Then we evaluate the risk based on the total score P to get the speed of information update and set an evaluation plan for a single information area.

Table 4 Fire risk evaluation system

Evaluation Level	Score Interval	Information Update Interval
Extra High	$16<P\leq 20$	20min
High	$12<P\leq 16$	30min
General	$8<P\leq 12$	45min
Low	$4<P\leq 8$	60min
Extra Low	$0\leq P\leq 4$	90min

2.5. Optimal combination

Through the above analysis and calculation, our decision-making model is determined to be composed of a basic part and an adjustment part. First of all, we need to divide several information unit areas for the entire jungle. Then, we define the area of each information unit as *S*, the speed of information update as *V*, cruise path of radio repeater drone as L, average fight speed of radio repeater drone as *v*, and the size and frequency of fire as respectively α and β. Therefore, the number of radio repeater drones can be calculated by the following formula:

$$a = \frac{L/v}{V}, V = \phi(\alpha, \beta) \qquad (4)$$

As SSA drones require to cover the entire information unit, we define the area that each SSA drone can detect as *s*, then the numbers of SSA drones are $b=S/s$.

From the above calculation, we can get a combination of drones in an information unit: radio repeater drones and b SSA drones. Then, the overall number of drones are aS_{all}/S radio repeater drones and bS_{all}/S SSA drones.

According to the distribution of Victoria Jungle, we divide it into 18 information units. Based on the above model, we assume that the observation coverage of each SSA drone is 25 square kilometers, and the average flight speed of the radio repeater drone is 60km/h. Then, the best combination of drones of each unit is listed in Table 5.

Table 5 Expenditure of each unit's drones combination

Information Unit Numbers	SSA Drones Numbers	Radio Repeater Drones Numbers	Cost ($)
1	69	6	750000
2	70	7	770000
3	67	3	700000
4	72	15	870000
5	70	5	750000
6	68	4	720000
7	69	5	740000
8	71	10	810000
9	72	15	870000
10	73	15	880000
11	70	10	800000
12	68	3	710000
13	68	5	730000
14	70	7	770000
15	69	6	750000
16	71	10	810000
17	73	15	880000
18	70	10	800000

Overall, to balance economic and safety factors and terrain[5], fire size, frequency, and other factors, we select the number and mix of SSA drones and radio repeater drones listed in Table 6.

Table 6 Optimal numbers and mix of SSA drones and radio repeater drones

	SSA Drones Numbers	Radio Repeater Drones Numbers	Cost ($)
Total	1260	151	14110000

3. Conclusion

We use Ant Colony Optimization to compute the shortest path, which can respond to the change of fire point. The more heuristic information, the stronger the reliability of the model, and the reference data comes from the satellite system. The source is relatively reliable.

We ignore some factors that may change the terrain, such as earthquakes, volcanic eruptions, etc., which may change the terrain and cause the model to be less accurate.

Since satellite instruments are far away from the ground, there is a little error in the acquisition of flame radiation power data, which may make the data inaccurate.

References

[1] Zhang Wenwen, Wang Qiuhua, Wang Ruichen, Long Tengteng, Wei Jianheng, Gao Zhongliang. Research Progress on Eucalyptus Forest Fire in Australia[J/OL]. World Forestry Research: 1-10[2021-02-07]. https://doi.org/10.13348/j.cnki.sjlyyj.2021.0004.y.

[2] Cheng Qing, Zhao Yuan, Li Yan-dong, Wu Han-yu, Hu Miao-miao. Research and Simulation Analysis of Urban UAV Track Algorithm [J]. Aeronautical Computing Technique, 2021, 51(01): 5-8+12.

[3] Wang Yongqing, HuoShuyan. Public Forest Management in Victoria, Australia [J]. World Forestry Research, 1999(03): 63-67.
[4] Xu Xiang. Path planning of forest fire warning UAV based on improved ant colony algorithm [J]. Electronic Quality, 2021(11): 50-55.
[5] Zhou Baoxiang. The influence of terrain features on UHF TV signal [J]. Electronic Technology, 1987(06): 28-30+46.

The Identification of Key Nodes in Aviation Networks

Bin Chen[1,2], **Yingming Wang**[1]

[1] Tencent Cloud Big Data College, Ma'anshan University, Ma'anshan, CHINA
postchenbin@163.com

Abstract. Identifying key nodes of aviation networks is a challenging and valuable issue, which depends on the measurement and ranking of node importance. Node importance can be measured statically according to certain network topology characteristic. The efficiency of this method is high. Node importance can also be measured dynamically according to certain network dynamic process. The result measured by this method is more accurate and reasonable. The current situation of this research is systematically described. The future trend of this research is revealed as well.

1. Introduction

Complex networks are helpful to understand the structural characteristics and behaviour laws of complex systems. The structure of a complex system can be modelled as a complex network with many nodes and edges. Structure affects behaviour. Behaviour changes structure in turn. The interaction between the network structure and the dynamic process is called "coevolution". Except for statistical models and topological structure characteristics, much attention has been paid on dynamic processes of complex networks in political, economic, biological and social fields.

Aviation networks are complex systems. Regardless of the specific locations of airports and the specific shapes of airlines, an aviation network can be modelled as a complex network, which takes airports as nodes and airlines between airports as edges [1]. How to quickly and accurately quantify the centrality of airports or airport sets in aviation networks has become a very challenging issue. Based on key nodes identification, the topology of an airport network can be optimized, robustness and fault tolerance of the network can be improved, path optimization algorithm can be designed, and delay of network transport can be reduced. In addition, the identification of key nodes in aviation networks also provides a new perspective for expressing, understanding and designing the properties and behaviours of neural networks because of the similarity between them.

2. Related works

The topology structure of complex systems can be mainly abstracted into two different kinds of complex networks. Scale-free complex networks are inhomogeneous, in which a few nodes have a large number of links but most nodes have a small number of links. Small-world complex networks are homogeneous, in which the majority of nodes have approximately the same number of links. For an aviation network with a large number of airports, "what impact will the removal of 5% of nodes have on network performance?" is a very meaningful problem. Most aviation networks are scale-free, in which the nodes are strongly inhomogeneous. It shows a power-law connectivity distribution. They have the dual characteristics of robustness and vulnerability [2]: they are robust under random attacks but vulnerable under deliberate attacks. In addition, cascading failures often occur in aviation networks. Although the node failure starts on a very small scale, the chain reaction is generated through the connections between nodes, and rapidly spreads to a large scale. According to empirical research and statistics, as long as 5% of key nodes fail at the same time, the entire aviation network may collapse [3].

The identification of key nodes depends on the measurement and ranking of node importance (or node centrality). So far, to effectively identify key nodes in aviation networks, many node importance measurement methods have been developed. These methods can be divided into two types. The social network analysis regards node importance as its "distinctiveness". In this way, node importance is generally measured according to certain network topology characteristic. The system science analysis regards node importance as its "destructiveness". In this way, node importance is generally measured according to certain network dynamic process.

3. The different methods for identifying key nodes

The "distinctiveness" of a node can be measured by an index about certain network topology characteristic. The index is either global or local. For example, degree is a local index, which describes the number of endpoints that a node act as. So, the degree can be used to measure the local influence range of a node. Closeness is a global index, which describes the average distance from a node to all the other nodes in the network. So, the closeness can be used to measure the centrality of a node in the network. Betweenness describes the number of the shortest paths through a node in the network. The larger the betweenness is, the greater the global influence of the node will be. Although the cluster coefficient is small, some nodes have a great influence on the information propagation. If they fail, structural holes will come into being. Based on this finding, node importance can be measured by the constraint coefficient of its structural hole. The ranking results of node importance measured by different indexes may be inconsistent. To reach a final comprehensive result, a new way for node importance measurement and ranking based on relative entropy was proposed in the research [4]. A multi index system was established for node importance measurement. Under this system, different results measured by various indexes were transformed into a discrete distribution about node importance. Then, an optimal solution for comprehensive measurement was found by means of linear programming.

There are mainly two aspects involved in node importance measurement based on "destructiveness". Firstly, the dynamic process model should be established for an aviation network. Secondly, the destructiveness caused by node failure should be measured. In aviation networks, a node and its edges will be removed from the network after the node fails. The remaining nodes and their edges that still work normally constitute a remaining network. The destructiveness of a node can be measured by the number of failed nodes or the interconnectedness of the remaining network. The indexes about the interconnectedness mainly include the number of subgraphs, the size of the largest subgraph, the number of unconnected node pairs and the average distance. It has been found that these typical indexes were often not universal when measuring various topologies of aviation networks. The indexes can only be applied to some topologies of networks. But when they are used to measure other topologies of networks, unreasonable results may be gotten. The main reason lies in the measurement methods. For the traditional methods, node importance is measured according to the change in the network performance before and after the node failure. As a result, few indexes can measure various topologies of networks effectively. In the research [5], node importance is measured by the comprehensive length of the remaining network. The comprehensive length is a new index, in which the infinite distance between an unconnected node pair is represented by the network size. The remaining network includes not only the nodes that still work normally but also the cascading failed nodes that have lost their connections with their neighbour nodes. When the comprehensive length is calculated, the distances between connected node pairs and the distances between unconnected node pairs are both counted in. This index is effective for various topologies of networks. More often, the reason why the key nodes are important in aviation networks is that they can cause large-scale cascading failure. The cascading failure was first proposed by Newman on power networks. Then, this phenomenon was found on other networks such as aviation networks. For various complex networks, typical cascading failure dynamic process models include the load capacity model [6], sandpile model [7], cascade model [8] and Optimal Power-flow Approach (OPA) model [9]. Among these models, the load capacity model is widely researched and applied to aviation networks. There are dynamic flows in aviation networks. If a node fails, its load will redistribute through its edges. When the loads of its neighbour nodes exceed their capacities, they will fail too. This is an iterative process. In previous researches, the effect of cascading failure is generally measured by the number of the failed nodes. However, aviation networks not only care about the scale of cascading failure, but also care about whether the remaining network can continue to work normally. At this time, the effect of cascading failure is not only related to the number of the cascading failed nodes, but also the interconnectedness of the remaining network.

4. Case study

To demonstrate the processes and compare the results of different measurement methods, an application case is studied here. The topology of an aviation network can be designed according to Kite network [10]. As shown in Figure 1, the Kite network consists of 10 nodes and 18 links.

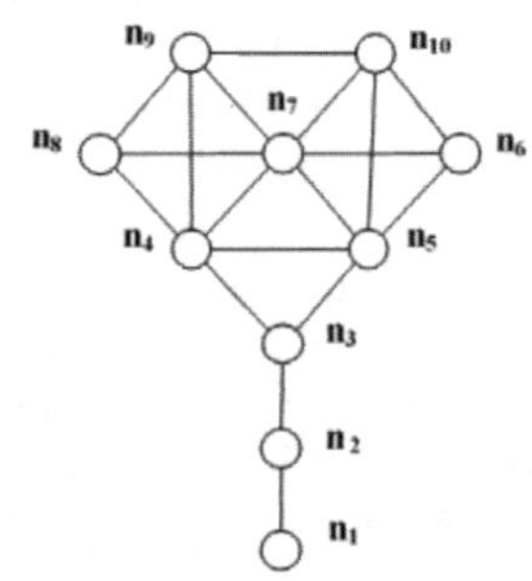

Figure 1. Kite network model.

Firstly, node importance can be measured statically by various indexes about the network topology characteristics. The values for each node measured by various indexes are listed in Table 1.

Table 1. Results of node importance measurement according to the network topology characteristics.

Node	n_1	n_2	n_3	n_4	n_5	n_6	n_7	n_8	n_9	n_{10}
Degree	1	2	3	5	5	3	6	3	4	4
Closeness	0.034	0.047	0.067	0.071	0.071	0.056	0.067	0.056	0.059	0.059
Betweenness	0	16	28	16.67	16.67	0	7.33	0	1.67	1.67
Constraint coefficient	1.25	0.556	0.494	0.47	0.47	0.706	0.475	0.706	0.578	0.578

According to Table 1, the ranking results are as follow.

- The result measured by degree is: $n_7 > n_4 = n_5 > n_9 = n_{10} > n_3 = n_6 = n_8 > n_2 > n_1$;
- The result measured by closeness is: $n_4 = n_5 > n_3 = n_7 > n_9 = n_{10} > n_6 = n_8 > n_2 > n_1$;
- The result measured by betweenness is: $n_3 > n_4 = n_5 > n_2 > n_7 > n_9 = n_{10} > n_6 = n_8 = n_1$;
- The result measured by constraint coefficient is: $n_4 = n_5 > n_7 > n_3 > n_2 > n_9 = n_{10} > n_6 = n_8 > n_1$.

If the degree accounts, n_7 is the most important because the value of n_7 is the biggest. If the closeness accounts, n_4 and n_5 are the most important. If the betweenness accounts, n_3 is the most important. If node importance is measured based on the structural holes, n_4 and n_5 are the most important because they have the smallest network constraint coefficient. Obviously, the ranking results are inconsistent. Because n_1 is an edge node in the network topology shown in Figure 1, it is easy to judge that it should be the most trivial. Then, the result measured according to betweenness is not accurate. And n_3 should be different from n_6 and n_8 because it is a bridge node. Then, the result measured by degree is not accurate. Moreover, n_3 should be different from n_7. Then, the result measured by the closeness is not accurate. Comparatively, the result measured by network constraint coefficient is more reasonable for Kite network.

Secondly, node importance can be measured by various indexes about the interconnectedness of remaining network after a node fails without considering dynamic process. The values for each node measured by various indexes are listed in Table 2.

Table 2. Results of node importance measurement according to the interconnectedness of the remaining network.

Node	n_1	n_2	n_3	n_4	n_5	n_6	n_7	n_8	n_9	n_{10}
Number of subgraphs	1	2	2	1	1	1	1	1	1	1
Size of the largest subgraph	0.9	0.8	0	0.4	0.4	0.9	0.5	0.9	0.8	0.8
Number of unconnected node pairs	0	8	14	0	0	0	0	0	0	0
Average distance	1.67	1.11	0.81	2.25	2.25	1.97	2.08	1.97	2	2

According to Table 2, the ranking results are as follow.

- The result measured by the number of subgraphs is: $n_2=n_3>n_1=n_4=n_5=n_6=n_7=n_8=n_9=n_{10}$;
- The result measured by the size of the largest subgraph is: $n_3>n_2>n_1=n_4=n_5=n_6>n_7=n_8=n_9=n_{10}$;
- The result measured by the number of unconnected node pairs is: $n_3>n_2>n_1=n_4=n_5=n_6=n_7=n_8=n_9=n_{10}$;
- The result measured by the average distance is: $n_4=n_5>n_7>n_9=n_{10}>n_6=n_8>n_1>n_2>n_3$.

If the number of subgraphs accounts, it is hard to distinguish the differences among most of nodes. Similarly, inaccurate results will be gotten if the size of the largest subgraph or the number of unconnected node pairs accounts. Moreover, if the average distance accounts, n_3 becomes the most trivial. Obviously, it is not accurate. Since n_1 is an edge node, it should be the most trivial.

Finally, node importance can be measured by various indexes about the effect of cascading failure. The load capacity model is the cascading failure dynamic process model of aviation networks. The initial load and capacity of each node in the Kite network are listed in Table 3.

Table 3. Initial load and capacity of each node in Kite network.

Node	n_1	n_2	n_3	n_4	n_5	n_6	n_7	n_8	n_9	n_{10}
Initial load	0	0.178	0.311	0.185	0.185	0	0.081	0	0.019	0.019
Capacity	0	0.21	0.373	0.222	0.222	0	0.098	0	0.022	0.022

The cascading failure process of each node is listed in Table 4. And the values for each initial failed node measured by various indexes are listed in Table 5.

Table 4. Cascading failure process of each node.

Node	**Cascading failure process**
n_1	n_1
n_2	$n_2 \rightarrow n_1$
n_3	$n_3 \rightarrow n_2 \rightarrow n_1$; $n_3 \rightarrow n_4 \rightarrow n_8$; $n_3 \rightarrow n_5 \rightarrow n_6$; $n_3 \rightarrow n_4 \rightarrow n_7 \rightarrow n_9$; $n_3 \rightarrow n_4 \rightarrow n_7 \rightarrow n_{10}$
n_4	$n_4 \rightarrow n_8$; $n_4 \rightarrow n_7 \rightarrow n_6$; $n_4 \rightarrow n_7 \rightarrow n_9$; $n_4 \rightarrow n_7 \rightarrow n_{10}$
n_5	$n_5 \rightarrow n_6$; $n_5 \rightarrow n_7 \rightarrow n_8$; $n_5 \rightarrow n_7 \rightarrow n_9$; $n_5 \rightarrow n_7 \rightarrow n_{10}$
n_6	n_6
n_7	$n_7 \rightarrow n_8$; $n_7 \rightarrow n_6$; $n_7 \rightarrow n_9$; $n_7 \rightarrow n_{10}$
n_8	n_8
n_9	$n_9 \rightarrow n_8$
n_{10}	$n_{10} \rightarrow n_6$

Table 5. Results of cascading failure.

Node	n_1	n_2	n_3	n_4	n_5	n_6	n_7	n_8	n_9	n_{10}
Number of failed nodes	0.1	0.2	1	0.6	0.6	0.1	0.5	0.1	0.2	0.2
Comprehensive length	3.33	4.66	10	8.89	8.89	3.58	8.15	3.58	5.02	5.02

According to Table 5, the ranking results are as follow.

- The result measured by the number of failed nodes is: $n_3 > n_4 = n_5 > n_7 > n_2 = n_9 = n_{10} > n_1 = n_6 = n_8$;
- The result measured by the comprehensive length is: $n_3 > n_4 = n_5 > n_7 > n_9 = n_{10} > n_2 > n_6 = n_8 > n_1$.

If the number of failed nodes accounts, the importance of n_1, n_6 and n_8 should be equal. But it is obviously not accurate. Because n_1 is an edge node, it should be the most trivial. Comparatively, the result measured by comprehensive length is more reasonable for Kite network.

In total, the result of node importance measured dynamically according to certain network dynamic process is more reasonable than that measured statically according to certain network topology characteristic.

5. Development trend

For the identification of key nodes in aviation networks, the development trend can be summarized as follows.

- The method of node importance measurement based on topology characteristics will be integrated with the method based on dynamic process model. To improve the effect of node importance measurement, appropriate topological characteristics should be found as the heuristic information of the dynamic process model.
- Machine learning algorithms will be integrated with the identification of key nodes in aviation networks. To improve the effect of node importance measurement, the important data about aviation networks such as the topology characteristics can be mined by machine learning algorithms.
- Big data technology will be integrated with the identification of key nodes in aviation networks. To improve the efficiency of node importance measurement, big data technology such as graph calculation mode and graph storage mechanism can be used for storing and processing of large-scale data in aviation networks.

6. Summary

The methods for measuring node importance of aviation networks can be categorized into two types. The social network analysis regards node importance as its "distinctiveness", while the system science analysis regards node importance as its "destructiveness". The "distinctiveness" of a node can be measured statically according to certain network topology characteristic, while the "destructiveness" of a node can be measured dynamically according to certain network dynamic process. Although the efficiency of the former is high, the experiment has showed that the measurement result is not as accurate as the result measured by the latter. In the future, the efficiency of node importance measurement algorithm can be improved by taking the network topology characteristic as the heuristic information of the dynamic process model. The efficiency and effect of node importance measurement for large-scale aviation networks can be improved by introducing machine learning algorithm and big data technology into the identification of key nodes.

Acknowledgments

The work is supported by University natural science fund project of China in Anhui province (KJ2020A0842), key project of academic support for top talents in university disciplines in Anhui province (gxbjZD2021095) and university key project of provincial quality engineering in Anhui province (2020jyxm1903).

References

[1] Hu J, Wang Y, He X, et al. Analysis and Application of Global Aviation Network Structure Based on Complex Network. *Computer Science*,2021,48(6):321-325.
[2] Vespignani A. Complex networks: the fragility of interdependency. *Nature*,2010,464(7291):984-985.
[3] Lai Y C, Motter A E, Nishikawa T. Attacks and cascades in complex networks. *Lecture Notes in Physics*,2004,650:299-310.
[4] Chen B, Wang Z X, Luo C. Integrated evaluation approach for node importance of complex networks based on relative entropy. *Journal of Systems Engineering and Electronics*, 2016, 27(6): 1219-1226.
[5] Chen B, Wu J P. A new approach for node centrality evaluation based on residual network performance measurement. *Concurrency and Computation Practice and Experience*, 2021, 33(16): e6560.
[6] Motter A E, Lai Y C. Cascade-based attacks on complex networks. *Physical Review E*, 2002,66(6):065-068.
[7] Lee D S, Goh K I, Kahng B. Sandpile avalanche dynamics on scale-free networks. *Physica A*,2004,338:84-91.
[8] Sansavini G, Hajj M R, Puri I K. A deterministic representation of cascade spreading in complex networks. *Europhysics Letters*,2009,87:48004.
[9] Carreras B A, Lynch V E, Dobson I. Critical points and transitions in an electric power transmission model for cascading failure blackouts. *Chaos*,2002,12(4):985-994.
[10] Yu H, Liu Z, Li Y J. Key nodes in complex networks identified by multi-attribute decision-making method. *Acta Physica Sinica*, 2013, 62(2): 54-62.

Forecasting and analysis of PM2.5 concentration in Hainan Province based on LSTM Neural Networks

Shuning Hu[1a*], Bing Liu[1, 2b*], Jizhen Wang[3c], Youchao Zhang[3d]

[1]College of Geodesy and Geomatics, Shandong University of Science and Technology, Shandong 266590, China

[2]College of Computer Science and Engineering, Shandong University of Science and Technology, Shandong 266590, China

[3]Shandong GEO-Surveying&Mapping Institute, Shandong 2500II, China

[a*]1902189408@qq.com, [b*]lb@sdust.edu.com, [c]793757540@qq.com, [d]819185484@qq.com

Abstract—The change of PM2.5 concentration not only reflects the change of air quality level, but also affects human life and health. However, because the PM2.5 change is caused by wide-ranging factors and the occurrence mechanism is complex, it is difficult to build a stable forecasting model with general physical methods. This study builds three different LSTM models to forecast the PM2.5 concentration of Hainan Province from 2019 to 2021. Model 1 is a fully connected neural network, which is also the base model. Model 2 is a single unit LSTM model. Model 3 is a multi-LSTM layer model. The conclusion indicate that the multi-layer LSTM model has the best forecasting precison for PM2.5 concentration, and the single unit LSTM model quite the opposite. The length of the time series has certain influence on the precision of PM2.5 forecasted by LSTM models, suggesting that a suitable time series length should be selected when building LSTM models.

1. Introduction

The frequent occurrence of hazy weather caused by Atmospheric Particles has made the research on PM2.5 concentration a hot topic. However, because of the non-linear relation between PM2.5 change and external influence, it is hard to forecast PM2.5 concentration by physical methods. ARIMA, which is a linear method, usually be used to build time series models, cannot capture the complex non-linear relationships in PM2.5 data.[1] Therefore, statistical forecasting methods commonly used are machine learning models and artificial neural network models, but the generalization ability of these models is closely related to the quality and quantity of samples.

Nowadays, machine learning models and neural network models have moved towards deep learning.For example, Ayturan et al.[2] used a deep learning approach to study the short-term change of PM2.5, and obtained relatively accurate prediction results within three hours after the current time. And a series of related studies for time series data have introduced to recurrent neural networks. LSTM is one of the representative methods, which has been applied to air quality forecasting on time series data with good results [3]. Vidushi et al. [4] proposed a multi-layer LSTM model to predict future atmospheric pollution condition; Weber et al. [5] constructed two forecasting models, a multi-level LSTM and an Encoder-Decoder, the outcomes of the experiment proved that the LSTM model outperformed the Encoder-Decoder framework. Chiou et al. [6] structured a deep neural network model by using historical hourly precipitation and consindering factors of wind, and first performed multiple one-dimensional convolutional reasoning, secondly get the results enter LSTM to forecast the change of PM2.5. Ping et al. [7] combined LSTM models with feature space correlation for predicting PM2.5 concentration. Zhao et al. [8] advanced a LSTM-FC fusion method which can effectively assess the future PM2.5 change of an air pollution station for 48h. Cong W et al. [9] raised a spatio-temporal convolutional LSTM neural network, in which the PM2.5 data of the present station and the nearby station were imported to the model after one-dimensional convolutional calculating.

Therefore, this study build three different LSTM models based on meteorological data and PM2.5 data to forecast the PM2.5 concentration of Hainan Province from 2019 to 2021 in the time series dimension and comparing their forecast accuracy by evaluation indexes.

2. Method

2.1. Data Preprocessing

PM2.5 data and meteorological data of Hainan Province from 2019 to 2021 are selected to support this experiment. The data cleaning work is first performed, and then we organize the PM2.5 data according to the temporal resolution of the meteorological data to align the data from two parts in terms of time dimension.

After completing the data collection and organization, the data are reorganized in the following three aspects:

(1) Time series data construction

Assuming that the time series length is n, the independent variables for each time series sample are the meteorological data and PM2.5 value of the previous n days, and the dependent variable is the PM2.5 value of the (n + 1)-th day.

(2) Trained dataset and tested dataset division

After all the samples of the time series data are obtained, take the former 70% of the samples to become trained dataset and the behind 30% of the samples are select to be tested dataset according to the proportion of 20% of the test samples. Next, we randomly shuffle the training sample data.

(3) Data standardization

Get new values by computing the average value and standard deviation from the trained samples, then apply this mean and std value to the trained samples and tested samples.

$$x_{new} = \frac{x - mean}{std} \tag{1}$$

2.2. LSTM

When building RNN models, it is easy to encounter the phenomenon of gradient dispersion and explosion, in order to improve this situation, Hochreiter and Schmidhuber [10] in 1997 put forward an advanced model: LSTM (namely Long Short-Term Memory), whose basic theory is the construction of the RNN. The advantage of LSTM lies in its specialized memory unit which can save previous information. The flowchart of the memory cell of LSTM is displayed in Fig.1. The previous information will be updated after they were processed by the following steps.

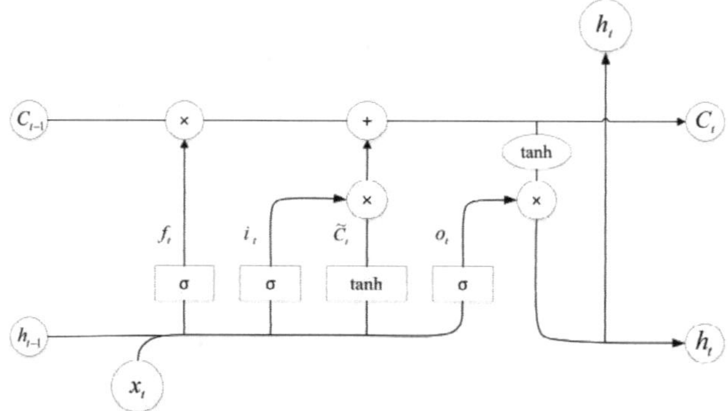

Fig. 1 The flowchart of the memory cell of LSTM

Each part of the flowchart of the LSTM memory cell are calculated according to formula (2)-formula (7) in turn:

$$f_t = \sigma\left(W_f[h_{t-1}, x_t] + b_f\right) \tag{2}$$

$$i_t = \sigma(W_i[h_{t-1}, x_t] + b_i) \tag{3}$$

$$\widetilde{C_t} = \varphi(W_C[h_{t-1}, x_t] + b_C) \tag{4}$$

$$o_t = \sigma(W_o[h_{t-1}, x_t] + b_o) \tag{5}$$

$$C_t = \tilde{C}_t \cdot (1 - f_t) + C_{t-1} \cdot f_t \tag{6}$$

$$h_t = \varphi(C_t) \cdot o_t \tag{7}$$

where x_t denotes the input of the vector at the current moment; W_f、W_i、W_C、W_o represent different weight matrices; $\mathrm{b_f}$、$\mathrm{b_i}$、$\mathrm{b_C}$、$\mathrm{b_o}$ are the biases of the corresponding weights. σ represent the sigmoid function; φ represent the tanh function; · denotes inter dot product. h_t is the hidden state vector with all useful information stored at the current moment as well as the previous moments.

2.3. Models Construction

(1) Fully connected neural network

Three layers are set in this model, including input section, output section and hidden section. Before the independent variable information enters the input section, it will be flattened to one dimension, where the hidden layer contain 32 units and the output layer contain 1 unit. The model loss function is MAE, a total of 200 epochs are trained, and the learning rate is adaptively reduced.

(2) LSTM models

First, a single unit LSTM model with 32 stacked layers is established, and the prediction results are output with a fully connected layer. In addition, a model with three LSTM units is established. Each unit has 32 stacked layers. At last, output the predictions using the fully connected layer.

These two LSTM models are both trained through mini-batch samples with mini-batch size 128, and they are both trained 200 epochs with learning rate reduced adaptively.

2.4. Models Evaluation

Based on the construction and training of models, we select RMSE and MAPE as the metrics to evaluate the models mentioned above in the test datasets. They are calculated as follow:

$$\mathrm{RMSE} = \sqrt{\frac{\sum_{j=1}^{n}\left(y_j - \hat{y}_j\right)^2}{n}} \tag{8}$$

$$\mathrm{MAPE} = \frac{1}{n}\sum_{j=1}^{n}\left|\frac{\hat{y}_j - y_j}{y_j}\right| \times 100\% \tag{9}$$

In the formula, n represents the size of test datasets; y_j and $\hat{y}_j$ individually represent the predicting value and the ground truth value of the j-th test samples.

3. Results and Analysis

This study construct and train three models, namely fully connected neural network, single LSTM unit model and LSTM with 3 LSTM units, corresponding to the following model 1, model 2, and model 3. When the time series lengths are set as 3 days, 7 days, and 15 days, the MAE value changes on the trained set and the tested set during model training are revealed in Fig.2 to Fig.3 respectively.

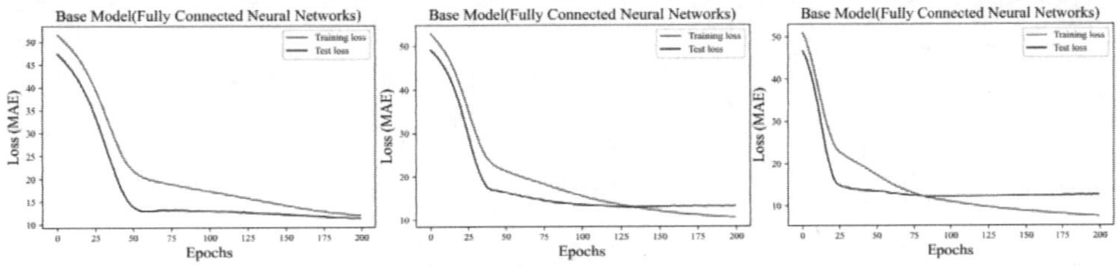

(a) series length is 3 (b) series length is 7 (c) series length is 15

Fig.2 The performance of model 1

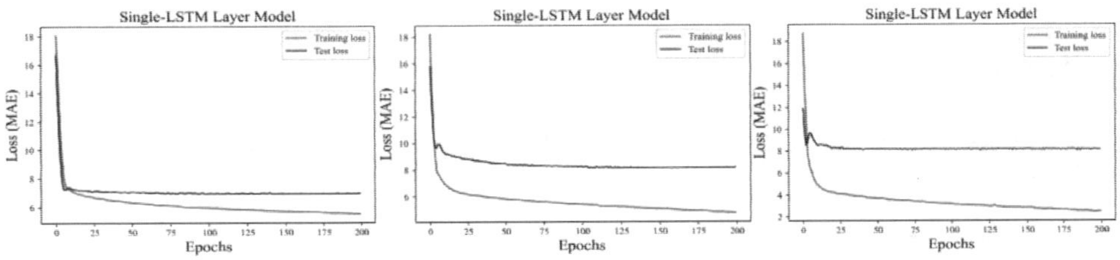

(a) series length is 3 (b) series length is 7 (c) series length is 15

Fig.3 The performance of model 2

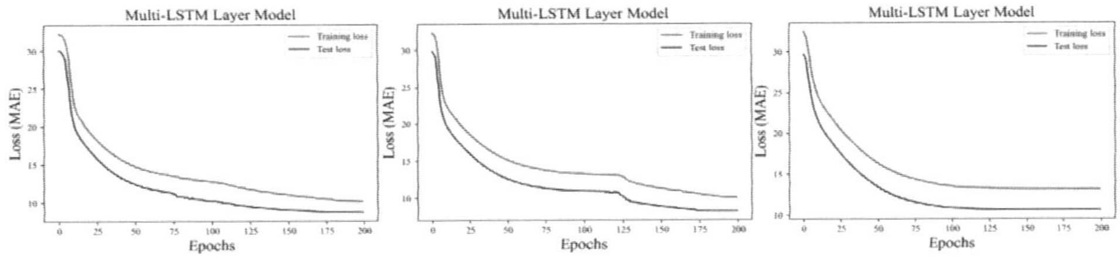

(a) series length is 3 (b) series length is 7 (c) series length is 15

Fig.4 The performance of model 3

Comparing Fig.3 and Fig.4, the single-LSTM layer model suffers from severe overfitting: starting from the 10th epoch, its loss function value (MAE) on the test dataset is much higher than that on the training dataset. Conversely, the multi-LSTM layer model is more stable and does not overfit throughout the 200 epochs of training. From Fig.2 to Fig.4, these figures can demonstrate that the prediction accuracy of multi-layer LSTM model is the highest.And followed by fully connected neural network, which also has good effect of forecasting the PM2.5 concentration, and the single-layer LSTM model has the worst performance. This is consistent with the conclusions reflected in Table 1.

Table 1. The RMSE and MAPE of each model with different time series length

series length	3 days		7 days		15 days	
index	RMSE	MAPE	RMSE	MAPE	RMSE	MAPE
Model 1	23.723	0.752	24.771	0.955	26.576	1.258
Model 2	33.059	702.947	32.691	710.253	32.727	312.893
Model 3	11.422	0.300	11.127	0.304	13.172	0.381
Average	22.735	234.666	22.863	237.171	24.158	104.844

Table 1 shows the RMSE and MAPE of each model with multiple time series lengths, from which we can draw that the forecast results of the multi-layer LSTM model perform the best. Although the length of the time series will have a certain impact on the precision of the model, this impact will not fundamentally alter the quality of each model. Besides, as shown in Table 1, the shorter the time series, the better the forecast results of the model. That is, when the time series length is 3 or 7, the overall function of the models improved significantly, compared to the function of the model whose time series length is 15 days.

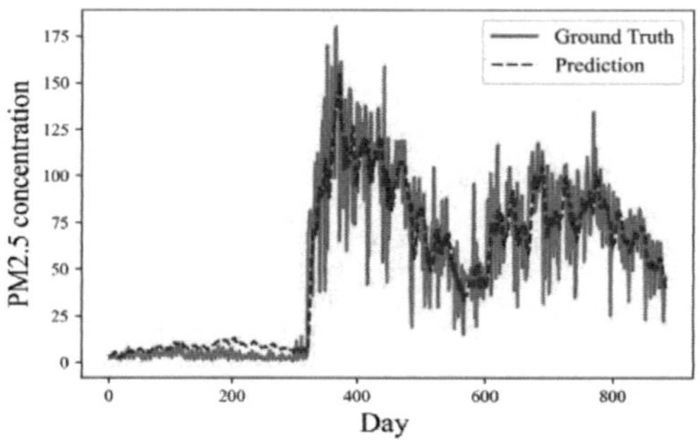

Fig.5 Hainan province 2019-2021 multi-layer LSTM PM2.5 forcasting

Fig. 5 is a graph which provide a poignant contrast between the ground truth data and the prediction data of PM2.5 concentration during air pollution in Hainan Province using the multi-layer LSTM model. This model can not only track the trend of PM2.5 concentration changes and respond to its fluctuation, but also realize relatively accurate prediction.

4. Conclusion

Based on the results and analysis presented above, the conclusions are obtained as below:

(1) LSTM neural networks can automatically learn the statistical probability distribution or non-linear relationship between meteorological data and PM2.5 concentration by computer, and then forecast the PM2.5 concentration in Hainan Province from 2019 to 2021, which can provide some reference for pollution early warning and air pollution prevention.

(2) RMSE and MAPE evaluation indicators are selected to access the forecast accuracy of LSTM models. The results reflect the multi-layer LSTM model has the best forecast accuracy for PM2.5concentration, and the single LSTM unit model quite the opposite.

(3) The length of time series has certain influence on the accuracy of PM2.5 forecasted by LSTM models, but it does not change the quality of each model. In addition, the shorter the time series, the better the overall prediction effect of the model.

References

[1] Ruby-Figueroa R., Saavedra J., Bahamonde N., et al. (2017) Permeate flux prediction in the ultrafiltration of fruit juices by ARIMA models. Journal of Membrane Science, 524: 108-116.
[2] Ayturan Y.A., Ayturan Z.C., Altun H.O., et al. (2020) Short-term prediction of PM2. 5 pollution with

deep learning methods. Global Nest Journal, 22(1): 126-131.
[3] Pak U., Kim C., Ryu U., et al. (2018) A hybrid model based on convolutional neural networks and long short-term memory for ozone concentration prediction. Air Quality, Atmosphere & Health, 11(8): 883-895.
[4] Weber S.A., Insaf T.Z., Hall E.S., et al. (2016) Assessing the impact of fine particulate matter (PM2. 5) on respiratory-cardiovascular chronic diseases in the New York City Metropolitan area using Hierarchical Bayesian Model estimates. Environmental Research, 151: 399-409.
[5] Chaudhary V., Deshbhratar A., Kumar V., et al. (2018) Time series based LSTM model to predict air pollutant's concentration for prominent cities in India. Environment Science, 5: 52260336.
[6] Huang C.J., Kuo P.H. (2018) A deep CNN-LSTM model for particulate matter (PM2. 5) forecasting in smart cities. Sensors, 18(7): 2220.
[7] Soh P.W., Chang J.W., Huang J.W. (2018) Adaptive deep learning-based air quality prediction model using the most relevant spatial-temporal relations. IEEE Access, 6: 38186-38199.
[8] Zhao J., Deng F., Cai Y., et al. (2019) Long short-term memory-Fully connected (LSTM-FC) neural network for PM2. 5 concentration prediction. Chemosphere, 220: 486-492.
[9] Wen C., Liu S., Yao X., et al. (2019) A novel spatiotemporal convolutional long short-term neural network for air pollution prediction. Science of the Total Environment, 654: 1091-1099.
[10] Hochreiter S., Schmidhuber J. (1997) Long short-term memory. Neural Computation, 9(8): 1735-1780.

AUTHOR INDEX